Electrical Impedance

Principles, Measurement, and Applications

Series in Sensors

Series Editors: Barry Jones and Haiying Huang

Other recent books in the series:

Electrical Impedance

Principles, Measurement, and Applications

Luca Callegaro

CRC Press
Taylor & Francis Group
Boca Raton London New York

CRC Press is an imprint of the
Taylor & Francis Group, an **informa** business

A TAYLOR & FRANCIS BOOK

CRC Press
Taylor & Francis Group
6000 Broken Sound Parkway NW, Suite 300
Boca Raton, FL 33487-2742

First issued in paperback 2016

© 2013 by Taylor & Francis Group, LLC
CRC Press is an imprint of Taylor & Francis Group, an Informa business

No claim to original U.S. Government works

Version Date: 20121207

ISBN 13: 978-1-138-19943-9 (pbk)
ISBN 13: 978-1-4398-4910-1 (hbk)

Visit the Taylor & Francis Web site at
http://www.taylorandfrancis.com

and the CRC Press Web site at
http://www.crcpress.com

Contents

Figures

Tables

Preface

...nec dubitamus multa esse quae et nos praeterierint;
homines enim sumus et occupati officiis.

[... nor do we doubt that many things have escaped us also; for
we are but human, and beset with duties.]

GAIUS PLINIUS SECUNDUS (Pliny the Elder)
Preface of *Naturalis Historia*, 77–79 AD.

Motivation The interest in the accurate measurement of electrical impedance is shared by scientists and engineers from different backgrounds. Impedance measurements can be performed on an impedance standard, to perform a calibration and issue a calibration certificate. Electromechanical appliances and electronic components can be characterized by impedance measurement to identify the parameters of their equivalent electrical model. Properties such as resistivity, permittivity, and permeability of material samples can be derived from impedance measurements conducted with proper electrical fixtures. Biological quantities related to a tissue, or even to a living being, can be related to their impedance. Sensors of many physical quantities can have electrical impedance as their output. *Impedance spectroscopy* permits to follow the evolution of an ongoing electrochemical reaction; *impedance tomography* is an imaging technique.

Despite such broad range of existing applications, and the potential for new ones, high school and university courses show a marginal interest to the subject of impedance measurement. Often, impedance measurement techniques are described as minor variations of the corresponding dc resistance measurements; recent publications may cite obsolete measurement techniques.

Difficulties When performing an impedance measurement, the experimenter faces conceptual and practical difficulties that are not encountered in resistance measurements. Voltages and currents become geometry-dependent quantities, and different parts of the measurement circuit can interact in unexpected ways because of mutual capacitances and inductances. Even commercial impedance meters ask for careful wiring techniques, which may involve a number of conductors. The measurement result can be expressed in a variety of representations, related by non-trivial mathematical transformations, prone to be misinterpreted.

Outline Chapter 1 recollects main definitions of the quantities related to impedance, some theorems of particular interest, and the issue of impedance representation. Chapter 2 introduces the problem of *impedance definition*, electromagnetic ways to distinguish the impedance to be measured from the environment. Chapter 3 gives a list of devices, appliances, circuits, and instruments employed as building blocks of impedance measurement setups. Chapter 4 attempts a classification of main impedance measurement methods, and for the most important give details on their implementation when a specific impedance definition is chosen. The increasing use of mixed-signal electronics in impedance measurement setups is discussed in Chapter 5.

Chapter 6 gives a list of applications and some details on the measurement of electromagnetic properties of materials.

Chapters 7 to 9 are devoted to impedance metrology. After Chapter 7, an introduction, Chapter 8 is devoted to artifact impedance standards, the material basis of measurement traceability. Chapter 9 deals with primary metrology: the realization and reproduction of SI impedance units.

Limitations The science of impedance measurement spans over more than 150 years, and even a condensed recollection of all important theoretical results, measurement methods and implementations is beyond the scope of the book and, frankly, of the author's capacity. A large part of circuits reported are principle schematics; equations expressing a measurement model are reported without an explicit derivation. No operative measurement procedures or troubleshooting techniques are reported. No hint about the expression of measurement uncertainty is given.

References and further reading As a partial compensation for the reader for the limitations listed above, every time this limitation was particularly apparent the author tried to include references to excellent papers reporting the pregnant details omitted in the book.

The choice of references included does not follow any systematic criterion, and no attempt for completeness has been pursued. Whenever possible, milestone or recent papers in English language, published on peer-reviewed journals, have been preferred over conference papers and technical notes. Just a bunch of references to review papers and books are present, because only a few have been published in the recent past. When a historical reference is given, it is usually the first one the author is aware of; the choice does not imply a serious historical research.

Acknowledgments I wish to thank my colleagues Giampiero Amato and Walter Bich; and Massimo Ortolano (Politecnico di Torino, Italy), for their critical reviewing of selected chapters. Giorgio Bertotti continuously encouraged me during the manuscript writing. I am indebted with my colleagues Stefano Borini, Cristina Cassiago, Natascia De Leo, Vincenzo D'Elia, Francesca Durbiano, Fausto Fiorillo, Matteo Fretto, and Umberto Pogliano; and with Alexandre Bounouh (LNE), John Fiander (NMIA), Jan Kučera (ČMI), and Jürgen Schurr (PTB), for kindly providing material included in the book.

The General Radio Historical Society, through Henry P. Hall, provided me with copies of several General Radio instrument manuals and *General Radio Experimenter* issues, and gave the permission to reprint some material in the book. Some images from the J. Res. Natl. Bur. Std. are reprinted with permission of NIST. Excerpts from guides in metrology are published with permission of the BIPM director.

Analog Devices Inc., USA, Huber+Suhner AG, Switzerland, and Fluke Corp., USA, provided me permission to use images from their manuals and datasheets. Philippe Roche, CNRS, gave me permission to reprint the reactance chart in Appendix D.

I thank Stefania for her patience and support, both in general and particularly during the preparation of this book.

1

Basics

CONTENTS

1.1   Two-terminal circuit elements

The definitions given in this chapter apply strictly only to *linear* and *time-invariant* electrical devices and circuits.[1] Most of the discussion is centered on *two-terminal* elements, also called *two-terminal networks*, or *one-ports*, see Fig. 1.1; a voltage $v(t)$ across the two terminals, and a current $i(t)$ entering one terminal and exiting the other can be defined. However, impedance metrology deals often with elements having more than two terminals, see Sec. 1.6.

Figure 1.1
A two-terminal element.

1.2   Resistors, capacitors, inductors

1.2.1   Resistors

A *resistor* is an element that obeys *Ohm's law*:

$$v(t) = R\,i(t) \quad\quad \text{or, equivalently,} \quad\quad i(t) = Gv(t), \quad \text{for all } t. \quad\quad (1.1)$$

R is the *resistance*, and $G = R^{-1}$ the *conductance*, of the element.
   Special cases of resistors are

the *open circuit*, defined by $G = 0$: that is, $i(t) \equiv 0$ for any $v(t)$;

the *short circuit*, defined by $R = 0$: that is, $v(t) \equiv 0$ for any $i(t)$.

---

[1] For a more general treatment, including nonlinear or time-variant circuit elements, see Chua et al. (1987).

Within the SI (see App. A), the unit of resistance is the ohm (Ω) and the unit of conductance is the siemens (S).

Resistors are perfectly *dissipative* elements: the electrical power

$$p(t) = v(t)i(t) = Ri^2(t) = Gv^2(t) \geq 0,$$

absorbed by the element is instantaneously dissipated.[2]

1.2.2 Capacitors and inductors

An ideal (or *pure*) *capacitor* is an element which electrical charge $q(t)$ and voltage $v(t)$ satisfy the relation

$$q(t) = Cv(t), \quad \text{for all } t. \tag{1.2}$$

C is the *capacitance* of the element. In SI, the unit of capacitance is the farad (F).

A (pure) *inductor* is an element in which magnetic flux $\Phi(t)$ and current $i(t)$ satisfy the relation

$$\Phi(t) = Li(t), \quad \text{for all } t. \tag{1.3}$$

L is the *inductance* of the element. In SI, the unit of inductance is the henry (H).[3]

Capacitors and inductors are *lossless* elements: the electrical power $p(t) = v(t)i(t)$ entering a capacitor or an inductor is not dissipated, but stored as *electrostatic energy* in capacitors,

$$E(t) = q(t)v(t) = \frac{1}{2}Cv^2(t) = \frac{1}{2}\frac{q^2(t)}{C},$$

or *magnetic energy* in inductors,

$$E(t) = \Phi(t)i(t) = \frac{1}{2}Li^2(t) = \frac{1}{2}\frac{\Phi^2(t)}{L}.$$

1.3 Phasors

Sinusoidal steady-state analysis deals with electrical circuits composed of linear, time-invariant elements driven by sinusoidal voltage or current source(s),

---

[2]Not necessarily into heat. R can be, for example, the *radiation resistance* of an antenna, where the input power is converted into power of radiant electromagnetic waves, see Sec. 1.8.

[3]In addition to resistors, capacitors, and inductors, a fourth two-terminal element called the *memristor* can be defined by the remaining relationship between q and Φ, see Chua (1971).

at a given frequency f (angular frequency $\omega = 2\pi f$), after the effect of transients have come to an end.

Given a generic sinusoidal voltage/current signal,

$$x(t) = X_{\max} \cos(\omega t + \varphi) = \text{Re}\left[X_{\max} \exp(j\omega t + \varphi)\right],$$

having a positive maximum amplitude $X_{\max}$, angular frequency ω, and phase φ, it is possible to associate to the signal a complex quantity

$$X = X_{\max} \exp(j\varphi),$$

which is the *phasor* of that signal. If ω is known, the phasor X completely identifies the signal $x(t)$. The correspondence $x(t) \Leftrightarrow X$ is also called *Steinmetz transform*.

The phasor X can be expressed in Cartesian form, by stating its real $\text{Re}\,X$ and imaginary $\text{Im}\,X$ parts, or in polar form, by stating its magnitude $|X|$ and phase φ_X, also often denoted as $\arg X$.

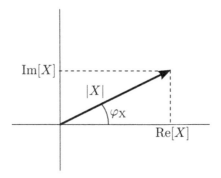

Figure 1.2
The representation of phasor X in the complex plane.

The standard graphical representation of a phasor is a vector in the complex plane, from the origin to the point X, see Fig. 1.2. Such representation is particularly useful when several phasors have to be considered, to easily identify their magnitude and phase relationships. Often, a single graph can combine phasors of different physical quantities (e.g., voltage and current), properly rescaled.

In general, circuit analysis ask for the solution of a system of differential equations. With the help of the concept of phasors, the sinusoidal steady-state analysis of linear, time-invariant electrical circuits can be reduced to the solution of a system of linear algebraic equations in the complex domain.

1.4 Impedance and admittance

Consider the circuit in Fig. 1.3(a), where the network $\mathcal{N}$ (composed of linear, time-invariant passive circuit elements) is driven by a current generator.

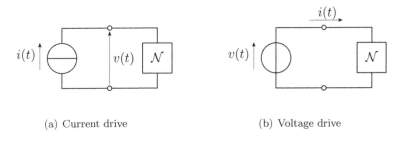

(a) Current drive   (b) Voltage drive

Figure 1.3
Network $\mathcal{N}$, driven by (a) current or (b) voltage sources.

Its input is a sinusoidal current source $i(t)$, with angular frequency ω:

$$i(t) = I_{\max} \cos(\omega t + \varphi_I),$$

to which a phasor

$$I = I_{\max} \exp(\mathrm{j}\varphi_I)$$

can be associated.

In the sinusoidal steady state, the voltage $v(t)$ is then also sinusoidal with angular frequency ω, to which a phasor V can be associated:

$$v(t) = V_{\max} \cos(\omega t + \varphi_V), \qquad V = V_{\max} \exp(\mathrm{j}\varphi_V);$$

the *(driving point) impedance* $Z(\omega)$ of network $\mathcal{N}$ at frequency ω can be defined as the ratio of phasors V and I,

$$Z(\omega) = \frac{V}{I}.$$

The SI unit of impedance is the ohm (Ω).

As a dual situation, consider the circuit in 1.3(b), where the same network $\mathcal{N}$ is driven by a sinusoidal voltage source $v(t)$, to which the phasor V is associated, and draws a current $i(t)$ having phasor I. We define the *(driving point) admittance* Y of network $\mathcal{N}$ at angular frequency ω the quantity

$$Y(\omega) = \frac{I}{V}.$$

The SI unit of admittance is the siemens (S). From the above definitions, it follows that $Z(\omega) = [Y(\omega)]^{-1}$.

Because of the linearity hypothesis, the impedance (admittance) of a network $\mathcal{N}$ is dependent on frequency ω but independent of phasor I (or V) of the source current (voltage) signal employed in the definition.

The calculation rules for expressing series and parallels of impedances and admittances follow the corresponding rules for series and parallels of resistances and conductances, respectively.

1.4.1 Quantities related to impedance

For a given network $\mathcal{N}$, Tab. 1.1 gives the definition of other quantities related to its impedance $Z(\omega)$.

Table 1.1
Quantities related to impedance.

Quantity	Expression	SI Unit
Resistance	$R = \operatorname{Re} Z$	Ω
Reactance	$X = \operatorname{Im} Z$	Ω
Conductance	$G = \operatorname{Re} Y$	S
Susceptance	$B = \operatorname{Im} Y$	S
Quality factor	$Q = \left\|\dfrac{\operatorname{Im} Z}{\operatorname{Re} Z}\right\| = \left\|\dfrac{\operatorname{Im} Y}{\operatorname{Re} Y}\right\|$	
Dissipation factor	$D = \left\|\dfrac{\operatorname{Re} Z}{\operatorname{Im} Z}\right\| = \left\|\dfrac{\operatorname{Re} Y}{\operatorname{Im} Y}\right\|$	
Time constant	$\tau = \dfrac{1}{\omega}\dfrac{\operatorname{Im} Z}{\operatorname{Re} Z} = -\dfrac{1}{\omega}\dfrac{\operatorname{Im} Y}{\operatorname{Re} Y}$	s

Note that, for a general network, $R \neq G^{-1}$ and $X \neq B^{-1}$. Q and D are defined as unsigned (positive) quantities; τ is a signed (positive or negative) quantity.

1.4.2 Impedance and admittance of pure elements

From definitions (1.1), (1.2), and (1.3), it follows that impedance and admittance of pure elements are real quantities for resistors, and pure imaginary for inductors and capacitors; at given frequency ω, impedances and admittances of pure two-terminal elements are given in Tab. 1.2.

1.4.3 Series and parallel representations

A two-terminal pure resistor and a pure reactive element (an inductor or a capacitor) can be connected in series, or in parallel, to obtain a new two-

Table 1.2
Impedance and admittance of pure elements.

Element	Symbol	Impedance	Admittance
Resistor	R	$Z = R$	$Y = \dfrac{1}{R}$
Capacitor	C	$Z = -\mathrm{j}\dfrac{1}{\omega C}$	$Y = \mathrm{j}\omega C$
Inductor	L	$Z = \mathrm{j}\omega L$	$Y = -\mathrm{j}\dfrac{1}{\omega L}$

terminal network. The four possible combinations, and their impedance (and related quantities), are listed in Tab. 1.3.

Given a generic passive and linear network $\mathcal{N}$, having at given angular frequency ω an impedance $Z(\omega)$ (and admittance $Y(\omega)$), it is always possible to find appropriate values for the elements of Tab. 1.3, such that *at the angular frequency ω* each network has the same $Z(\omega)$ of $\mathcal{N}$:

$$R_\mathrm{s} = \mathrm{Re}\, Z, \qquad L_\mathrm{s} = \omega^{-1}\, \mathrm{Im}\, Z, \qquad C_\mathrm{s} = -(\omega\, \mathrm{Im}\, Z)^{-1};$$
$$R_\mathrm{p} = (\mathrm{Re}\, Y)^{-1}, \qquad L_\mathrm{p} = -(\omega\, \mathrm{Im}\, Y)^{-1}, \qquad C_\mathrm{p} = \omega^{-1}\, \mathrm{Im}\, Y. \qquad (1.4)$$

Such combinations are are called the *equivalent series* or parallel representations of $\mathcal{N}$.

The expression of $Z(\omega)$ in terms of series and parallel representations is among the typical features of impedance meters, which implement in firmware equations like Eq. (1.4) and those reported in Tab. 1.3.

The convenience in choosing a particular series/parallel representation of $Z(\omega)$ is linked to the physical properties of the network considered, see also Sec. 1.9.

It is worth noting that

- some representations ask for – seemingly unphysical – negative inductance or capacitance values;

- a representation is *not* an electrical model of $\mathcal{N}$ valid for frequencies other than ω. However, a particular representation *can* be an appropriate model of a physical element in a limited frequency range. For example, the series $L_\mathrm{s} - R_\mathrm{s}$ representation can be an appropriate electrical model of the behavior of real inductors at low frequencies; the parallel $C_\mathrm{p} - R_\mathrm{p}$ representation can be an electrical model of real capacitors at audio frequencies.

Table 1.3

Series and parallel representations of a general network $\mathcal{N}$, with the mathematical expression of correspondent impedance and related quantities.

	Z	Y	Q	D	τ
R_s L_s	$Z = R_s + j\omega L_s$	$Y = \dfrac{R_s}{R_s^2 + \omega^2 L_s^2} - j\dfrac{\omega L_s}{R_s^2 + \omega^2 L_s^2}$	$Q = \dfrac{\omega L_s}{R_s}$	$D = \dfrac{R_s}{\omega L_s}$	$\tau = +\dfrac{L_s}{R_s}$
L_p R_p	$Z = \dfrac{\omega^2 R_p L_p^2}{R_p^2 + \omega^2 L_p^2} + j\dfrac{\omega R_p^2 L_p}{R_p^2 + \omega^2 L_p^2}$	$Y = \dfrac{1}{R_p} - j\dfrac{1}{\omega L_p}$	$Q = \dfrac{R_p}{\omega L_p}$	$D = \dfrac{\omega L_p}{R_p}$	$\tau = +\dfrac{R_p}{\omega^2 L_p}$
R_s C_s	$Z = R_s - j\dfrac{1}{\omega C_s}$	$Y = \dfrac{\omega^2 R_s C_s^2}{1 + \omega^2 R_s^2 C_s^2} + j\dfrac{\omega C_s}{1 + \omega^2 R_s^2 C_s^2}$	$Q = \dfrac{1}{\omega R_s C_s}$	$D = \omega R_s C_s$	$\tau = -\dfrac{1}{\omega^2 R_s C_s}$
C_p					

1.5   Power and RMS value

1.5.1   Power

The instantaneous power $p(t)$ entering a two-port network $\mathcal{N}$ having impedance $Z(\omega)$, in the sinusoidal steady state at angular frequency ω, is

$$
\begin{aligned}
p(t) &= v(t)\,i(t) \\
&= V_{\max}\cos(\omega t + \varphi_V) \cdot I_{\max}\cos(\omega t + \varphi_I) \\
&= \frac{1}{2}V_{\max}\,I_{\max}\left[\cos(\varphi_V - \varphi_I) + \cos(2\omega t + \varphi_V + \varphi_I)\right]
\end{aligned}
$$

which is the sum of a constant term, and one oscillating at frequency 2ω. The constant term, which is also the time average $\overline{p(t)} = \dfrac{1}{T}\displaystyle\int_0^T p(t)\mathrm{d}t$ of $p(t)$, is called the *active power* dissipated in $\mathcal{N}$:

$$
P \equiv \overline{p(t)} = \frac{1}{2}V_{\max}\,I_{\max}\left[\cos(\varphi_V - \varphi_I)\right]; \tag{1.5}
$$

$\psi = \cos(\varphi_V - \varphi_I)$ is the *power factor* of $\mathcal{N}$.

Expressing V and I as phasors[4] it is possible to define the *complex power* S as

$$
S \equiv \frac{1}{2}VI^*. \tag{1.6}
$$

From (1.6),

$$
P = \operatorname{Re} S = \operatorname{Re}\left[\frac{1}{2}VI^*\right] = \frac{1}{2}\operatorname{Re} Z\,|I|^2 = \frac{1}{2}\operatorname{Re} Y\,|V|^2.
$$

The imaginary part of S is the *reactive power*:

$$
Q \equiv \operatorname{Im} S = \operatorname{Im}\left[\frac{1}{2}VI^*\right] = \frac{1}{2}\operatorname{Im} Z\,|I|^2 = \frac{1}{2}\operatorname{Im} Y\,|V|^2
$$

and represents the power exchanged back and forth between the reactive part of Z and the generator.

1.5.2   RMS value and a notational ambiguity

In order to get rid of the factor $\frac{1}{2}$ in Eq. (1.5), (1.6), and following, it is common in electrical engineering to speak of *effective* or *root mean square* (RMS) values of sinusoidal voltage and currents as the phasors

$$
V_{\mathrm{RMS}} = \frac{1}{\sqrt{2}}V, \qquad I_{\mathrm{RMS}} = \frac{1}{\sqrt{2}}I.
$$

---

[4]The symbol I^* denotes the complex conjugate of I.

In this way, power expressions have the same form as in dc steady state: for example, for a pure resistor, $P = R I_{\text{RMS}}^2$.

Expressions involving *ratios* of voltage and current phasors, as impedance definitions, can be rewritten by using rms phasors instead of maximum amplitude phasors. Therefore, an ambiguous notation can be acceptable: the actual meaning of V and I in an expression must be deduced from the context.

RMS value definition is not limited to sinusoidal steady state. For any quantity $x(t)$, periodic in time[5] with period T (that is, $x(t) = x(t + kT)$ for any integer k) we can define its RMS value as

$$X_{\text{RMS}} = \sqrt{\overline{x^2(t)}} = \sqrt{\frac{1}{T} \int_\tau^{\tau+T} x^2(t)\, dt} \qquad \text{for any } \tau. \qquad (1.7)$$

1.6   Beyond two-terminal networks

Multi-terminal and multi-port network theory is the subject of entire treatises; only a short extract will be given here, and the reader is encouraged to refer to systematic treatments (e.g., Chua et al. (1987); Pozar (2005)). The following deals with passive, time-invariant, linear m-terminal networks in the sinusoidal steady state at frequency ω.

1.6.1   Multi-terminal networks, transimpedance

A graphical representation of an m-terminal network is shown in Fig. 1.4. An inward current J_k flows in the generic terminal k, which has a potential E_k relative to an arbitrary point O; the potential difference between two terminals is $V_{jk} = E_j - E_k$. On a m-terminal network, $m - 1$ voltages and currents can be independently specified (because of Kirchhoff's laws: $\sum_{k=1}^m J_k = 0$ and $V_{12} + V_{23} + \ldots + V_{m1} = 0$).

As a natural extension of the impedance definition given in Sec. 1.4, it is possible to introduce the concept of four terminal impedance, or *trans*impedance.

With reference to Fig. 1.5, we can define the four-terminal impedance $Z_{ij,kl}$ as the ratio of the open-circuit voltage phasor V_{ij} between terminals i and j, to the current phasor flowing into terminal k and out of terminal l, when all other terminals are open-circuited:

$$Z_{ij,kl} = \left. \frac{V_{ij}}{I_{kl}} \right|_{I_r=0;\ r \neq k,l} ;$$

---

[5]The definition can be extended to stationary random quantities $x(t)$ by taking an appropriate limit for $T \to \infty$.

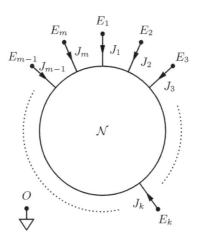

Figure 1.4
m-terminal network $\mathcal{N}$. To each terminal k a potential E_k (measured respect a reference potential E_O) and an inward current J_k can be associated.

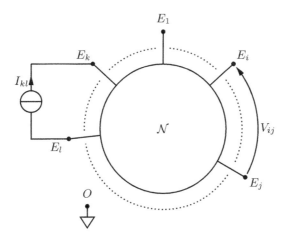

Figure 1.5
Definition of the four-terminal impedance $Z_{ij,kl}$ on network $\mathcal{N}$.

Impedances of the form $Z_{il,kl}$, i.e., with one terminal in common, are referred to as *three-terminal impedances*. Those of the form $Z_{kl,kl}$ are the two-terminal impedances already defined in Sec. 1.4.

Direct analysis of linear, passive m-terminal networks as circuit elements is possible with the *indefinite admittance* and *impedance* matrices and related theorems (Shekel, 1954; Zadeh, 1957; Sharpe, 1960). However, such analysis is less common than n-port network analysis, discussed in the following section.

1.6.2   Multi-port networks

It is very common to find that the terminals of an m-terminal network are naturally coupled in n pairs, called *ports*, such that the current entering one terminal of the pair is equal to the current exiting from the other terminal of the same pair.[6] Often, the physical port is actually a coaxial conductor pair (Secs. 2.2.1 and 3.5.2). The two-terminal network of Fig. 1.1 is a one-port network.

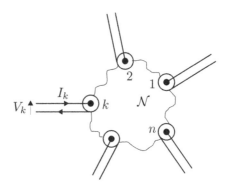

Figure 1.6
Generic n-port network; each port is here drawn as a coaxial pair.

For the linear n-port network of Fig. 1.6, at each port k we can measure the voltage V_k and the (inward) current I_k, which can be grouped in column[7] vectors $\boldsymbol{V} = [V_1, V_2, \ldots, V_n]^{\mathrm{T}}$ and $\boldsymbol{I} = [I_1, I_2, \ldots, I_n]^{\mathrm{T}}$.

1.6.3   Matrix descriptions of multi-port networks

---

[6]The port representation is so convenient that when such pairing is not possible, for example when the currents of the m terminals are all different, one terminal (often called the *ground*) is splitted between more than one port.

[7]$\boldsymbol{A}^{\mathrm{T}}$ stands for transposed vector or matrix.

1.6.3.1 Impedance matrix

By defining elements Z_{ij} as

$$Z_{ij} = \frac{V_i}{I_j}\bigg|_{I_k=0,\, k\neq j}$$

(the condition $I_k = 0$, $k \neq j$ means that all ports $k \neq j$ are open-circuited), it is possible to construct the *impedance matrix*

$$\mathbf{Z} = \begin{bmatrix} Z_{11} & Z_{12} & \cdots & Z_{1n} \\ Z_{21} & Z_{22} & \cdots & Z_{2n} \\ & & \ddots & \\ Z_{n1} & Z_{n2} & \cdots & Z_{nn} \end{bmatrix}.$$

Because of linearity,

$$\mathbf{V} = \mathbf{Z}\mathbf{I}; \qquad \begin{bmatrix} V_1 \\ V_2 \\ \cdots \\ V_n \end{bmatrix} = \begin{bmatrix} Z_{11} & Z_{12} & \cdots & Z_{1n} \\ Z_{21} & Z_{22} & \cdots & Z_{2n} \\ & & \ddots & \\ Z_{n1} & Z_{n2} & \cdots & Z_{nn} \end{bmatrix} \cdot \begin{bmatrix} I_1 \\ I_2 \\ \cdots \\ I_n \end{bmatrix}.$$

1.6.3.2 Admittance matrix

The admittance matrix can be similarly defined:

$$Y_{ij} = \frac{I_i}{V_j}\bigg|_{V_k=0,\, k\neq j};$$

the condition $V_k = 0$, $k \neq j$ means that all ports $k \neq j$ are short-circuited.

$$\mathbf{Y} = \begin{bmatrix} Y_{11} & Y_{12} & \cdots & Y_{1n} \\ Y_{21} & Y_{22} & \cdots & Y_{2n} \\ & & \ddots & \\ Y_{n1} & Y_{n2} & \cdots & Y_{nn} \end{bmatrix}$$

with the relation

$$\mathbf{I} = \mathbf{Y}\mathbf{V}.$$

1.6.3.3 Properties of impedance and admittance matrices

We have

$$\mathbf{Y} = \mathbf{Z}^{-1}.$$

Note that in general $Y_{ij} \neq Z_{ij}^{-1}$.

If the network is *reciprocal* matrices $\mathbf{Z}$ and $\mathbf{Y}$ are symmetric:

$$Z_{ij} = Z_{ji}; \qquad Y_{ij} = Y_{ji}.$$

1.6.3.4 Transmission matrix

An alternative representation of a two-port network is given by the *transmission matrix*, or $ABCD$ matrix, useful to analyze several networks connected in cascade:

$$\begin{bmatrix} V_1 \\ I_1 \end{bmatrix} = \begin{bmatrix} A & B \\ C & D \end{bmatrix} \begin{bmatrix} V_2 \\ -I_2 \end{bmatrix}.$$

The sign of I_2 takes into account the fact that, in practical applications, the current sign at port 2 is considered positive if current flows *outward* from the port.

1.6.3.5 Scattering parameter matrix

The *scattering parameter matrix* is a concept first introduced by Campbell and Foster (1920) and popularized in the 1950–60s (see, e.g., Matthews (1955), Carlin (1956), and especially Kurokawa (1965)). We give here the definition of small-signal scattering parameters.

A scattering parameter is defined in terms of a *reference impedance Z_0*. If Z_0 is real and equal for all network ports, the normalized incident a_i and reflected b_i wave amplitudes on port i are defined as

$$a_i = \frac{1}{2\sqrt{Z_0}} (V_i + Z_0 I_i), \qquad b_i = \frac{1}{2\sqrt{Z_0}} (V_i - Z_0 I_i). \qquad (1.8)$$

The meaning of definition (1.8) becomes more clear when inverted to find out V_i and I_i in terms of a_i and b_i,

$$V_i = \sqrt{Z_0} \, (a_i + b_i), \qquad I_i = \frac{1}{\sqrt{Z_0}} (a_i - b_i);$$

and compute the power entering port i,

$$P_i = \frac{1}{2} \, \text{Re}\,[V_i I_i^*] = \frac{1}{2} \left(|a_i|^2 - |b_i|^2 \right).$$

Scattering parameter matrix elements S_{ij} can now be defined as

$$S_{ij} = \frac{b_i}{a_i} \quad \text{when} \quad a_{k \neq i} = 0$$

that is, when no incident waves occur at other ports $k \neq i$, which happens when they are terminated with the reference impedance Z_0.

The scattering matrix permit to relate incident and reflected amplitudes: if $\boldsymbol{a} = [a_1, a_2, \ldots, a_n]^{\text{T}}$, and and $\boldsymbol{b} = [b_1, b_2, \ldots, b_n]^{\text{T}}$,

$$\boldsymbol{b} = \boldsymbol{S}\boldsymbol{a}; \qquad \begin{bmatrix} b_1 \\ b_2 \\ \ldots \\ b_n \end{bmatrix} = \begin{bmatrix} S_{11} & S_{12} & \ldots & S_{1n} \\ S_{21} & S_{22} & \ldots & S_{2n} \\ & & \ddots & \\ S_{n1} & S_{n2} & \ldots & S_{nn} \end{bmatrix} \cdot \begin{bmatrix} a_1 \\ a_2 \\ \ldots \\ a_n \end{bmatrix}$$

The definition of scattering parameters can be generalized to complex-valued reference impedances different for each port Kurokawa (1965).

1.6.3.6 Relationships among different matrix network representations

For non-degenerate networks, $\boldsymbol{Z}$, $\boldsymbol{Y}$, and $\boldsymbol{S}$ are related by the relations

$$\boldsymbol{Z} = \boldsymbol{Y}^{-1};$$

$$\boldsymbol{Z} = Z_0(\boldsymbol{I} + \boldsymbol{S})(\boldsymbol{I} - \boldsymbol{S})^{-1}, \tag{1.9}$$

where $\boldsymbol{I}$ is the identity matrix of order n.

Transmission matrix elements can be related to two-port impedance matrix with the following transformation:

$$Z_{11} = \frac{A}{C}; \qquad Z_{12} = \frac{AD - BC}{C}; \qquad Z_{21} = \frac{1}{C}; \qquad Z_{22} = \frac{D}{C}. \tag{1.10}$$

1.6.4 Two-port networks of particular interest

1.6.4.1 Mutual inductor

The *mutual inductor* is a two-port device governed by the equations

$$\Phi_1 = L_{11}\, i_1 + M\, i_2$$
$$\Phi_2 = M\, i_1 + L_{22}\, i_2$$

M is called the *mutual inductance* of the mutual inductor;[8] L_{11} and L_{22} are the two *self*-inductances. In the SI M, L_{11}, and L_{22} are measured in henry (H).

M can be positive, zero, or negative, and obeys the relation

$$M^2 \leq L_{11}L_{22}.$$

The *coupling coefficient*, k can be introduced:

$$k = \frac{M}{\sqrt{L_{11}\, L_{22}}}, \qquad |k| \leq 1.$$

In the sinusoidal steady state, we can write

$$V_1 = j\omega L_{11}\, I_1 + j\omega M\, I_2,$$
$$V_2 = j\omega M\, I_1 + j\omega L_{22}\, I_2,$$

and the *mutual reactance* $X_M = \omega M$ can be obviously introduced.

---

[8]The appearance of the same M in the two equations is due to the laws of electromagnetism, in particular Neumann's double integral expression.

1.6.4.2   Ideal transformer

The ideal transformer is a particular mutual inductor. Without loss of generality, one can write for a generic mutual inductor $L_{11} = t^2 L_{22}$, where t is called the *turns ratio*, see Sec. 3.3.2. If the mutual inductor has perfect coupling ($k = 1$), then $M = t L_{22}$. The inductor equations become

$$V_1 = j\omega L_{22} t (t I_1 + I_2),$$
$$V_2 = j\omega L_{22} (t I_1 + I_2);$$

hence,

$$\frac{V_1}{V_2} = t;$$

if, in addition, the reactances $\omega L_{11}, \omega L_{22}, \omega M \rightarrow \infty$, then

$$\frac{I_2}{I_1} = -t.$$

1.6.4.3   Gyrator

The *gyrator* is an ideal two-port element introduced by Tellegen (1948), which proposed also an electrical symbol, see Fig. 1.7(a). Its behavior is given by

$$\begin{bmatrix} v_1 \\ v_2 \end{bmatrix} = \begin{bmatrix} 0 & -R \\ R & 0 \end{bmatrix} \cdot \begin{bmatrix} i_1 \\ i_2 \end{bmatrix},$$

where the real parameter R is called the gyration resistance. Despite the presence of R in the impedance matrix, the gyrator is a lossless circuit element. It is linear but *nonreciprocal*.[9] Both classical (Mason et al., 1953) and quantum Hall effect (Sosso, 2001) devices can be electrically modeled as networks, including gyrators. If an impedance Z_2 is connected to port 2 of a gyrator

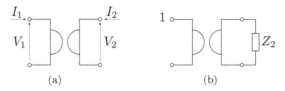

(a)   (b)

Figure 1.7
(a) The electrical symbol of a gyrator. (b) One-port network composed of a gyrator and an impedance Z_2.

___

[9]Consider a two-port network in two different working conditions: (a) voltage v_2 is generated by a source current i_1; (b) voltage v_1 is generated by a source current i_2. If $i_1 = i_2$ implies $v_1 = v_2$, the network is *reciprocal*. The gyrator is a nonreciprocal element (in fact, it is *anti-reciprocal*: $v_1 = -v_2$).

having gyration resistance R, see Fig. 1.7(b), the resulting one-port network has the impedance

$$Z_1 = \frac{R^2}{Z_2}.$$

In particular, if Z_2 is a capacitor C_2, then $Z_1 = j\omega \left(R^2 C_1\right)$, so the the behavior of port 1 is that of an inductor $L_1 = R^2 C_1$.

1.6.4.4 Transmission line

A *transmission line* is a two-port network that transmits electromagnetic energy from the input to the output port; it is a useful model of electrical connections, such as coaxial cables. The line can be characterized, for a given frequency (and propagation mode), by its equivalent series impedance per unit length z, and its equivalent parallel admittance per unit length y, and its length ℓ.

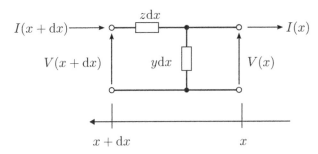

Figure 1.8
Infinitesimal section of a transmission line, modeled by Eq. (1.11).

A lumped equivalent circuit of an infinitesimal section, of length dx, of a transmission line is shown in Fig. 1.8. The propagation of voltage $V(x)$ and current $I(x)$ through length dx can be written as

$$V(x + dx) = V(x) + z\,dx\,I(x),$$
$$I(x + dx) = y\,dx\,V(x) + I(x). \tag{1.11}$$

The general solution of (1.11) is

$$V(x) = V_0^+ \exp(-\gamma x) + V_0^- \exp(+\gamma x),$$
$$I(x) = -\frac{1}{Z_0} V_0^+ \exp(-\gamma x) + \frac{1}{Z_0} V_0^- \exp(+\gamma x), \tag{1.12}$$

where $\gamma = \sqrt{zy}$ is the *propagation constant*, and $Z_0 = \sqrt{\dfrac{z}{y}}$ is the *characteristic impedance* of the transmission line.

V_0^+ and V_0^- are determined by boundary conditions at input and output ports. For given voltage $V(0)$ and current $I(0)$ for $x = 0$, Eq. (1.12) gives the celebrated *telegrapher's equations*

$$V(\ell) = \cosh(\gamma\ell)V(0) + Z_0 \sinh(\gamma\ell)I(0),$$

$$I(\ell) = \frac{1}{Z_0} \sinh(\gamma\ell)V(0) + \cosh(\gamma\ell)I(0), \tag{1.13}$$

which, for $\gamma\ell \ll 1$, can be approximated as

$$V(\ell) = \left(1 + \frac{(\gamma\ell)^2}{2}\right) V(0) + Z_0\gamma\ell I(0) \tag{1.14}$$

$$= \left(1 + zy\frac{\ell^2}{2}\right) V(0) + z\ell I(0),$$

$$I(\ell) = \frac{\gamma\ell}{Z_0}V(0) + \left(1 + \frac{(\gamma\ell)^2}{2}\right) I(0) \tag{1.15}$$

$$= y\ell V(0) + \left(1 + zy\frac{\ell^2}{2}\right) I(0).$$

Eq. (1.14) is the basis of *cable corrections*, see Sec. 2.4.

Labeling $x = \ell$ as port 1, $x = 0$ as port 2, Eq. (1.13) defines the transmission $ABCD$ matrix of the transmission line.

From Eq. (1.13), with transformation (1.10), the impedance matrix of the transmission line

$$\begin{bmatrix} V_1 \\ V_2 \end{bmatrix} = Z_0 \cdot \begin{bmatrix} \coth\gamma\ell & \operatorname{csch}\gamma\ell \\ \operatorname{csch}\gamma\ell & \coth\gamma\ell \end{bmatrix} \cdot \begin{bmatrix} I_1 \\ I_2 \end{bmatrix}$$

can be written. In the limit $\ell \to \infty$, the transmission line becomes a couple of matched loads, isolated from one another.

1.7 Impedance and linear response theory

The definition of impedance (Sec. 1.4) can be considered in the context of *linear response theory*.

Electrical quantities $i(t)$ and $v(t)$ are single-valued, real functions of the real quantity t. Although physically sound classical quantities should be continuous functions, it is generally accepted that *generalized functions* (or *distributions*) can enter the functional description of electrical quantities. For example, Dirac's delta function $\delta(t)$ can describe the idealized limit of fast events (such as voltage or current pulses) in which a detailed description is not interesting, or impossible, to achieve.

The theory of linear electrical circuits is traditionally developed with the Laplace transform, see, e.g., (Chua et al., 1987, Ch. 10). Here we propose an alternative but equivalent approach by using Fourier transform.[10]

1.7.1 Spectral analysis

Amplitude spectrum Given a real-valued, physical quantity $x(t)$, its *amplitude spectrum* $X(f)$ and $x(t)$ are related by the Fourier transform ($\mathcal{F}[\bullet]$) couple

$$X(f) = \mathcal{F}[x(t)] = \int_{\Re} x(t)\exp(-2\pi i\, ft)dt,$$

$$x(t) = \mathcal{F}^{-1}[X(f)] = \int_{\Re} X(f)\exp(2\pi i\, ft)df.$$

$X(f)$ has an internal symmetry[11] over f, so its description for $f \geq 0$ is sufficient.

If the unit of $x(t)$ is U, $[x(t)] = U$, then $[X(f)] = U\,s = U\,Hz^{-1}$.

Power spectral density Define the truncated signal $x_T(t)$ as

$$x_T(t) = \begin{cases} x(t) & \text{if} \quad |t| \leq T \\ 0 & \text{otherwise} \end{cases}$$

and take its amplitude spectrum $X_T(f) = \mathcal{F}[x_T(t)]$.

The power spectral density $S_X(f)$ is the limit, if existing,

$$S_X(f) = \lim_{T \to +\infty} \frac{1}{T}|X_T(f)|^2 = \lim_{T \to +\infty} \frac{1}{T}X_T(f) \cdot X_T^*(f) \qquad U^2\,Hz^{-1}.$$

$S_X(f)$ is positive real, and $S_X(f) = S_X(-f)$. Therefore, often the *one-sided power spectral density* $S_X^{ss}(f)$, defined only for $f \geq 0$ can be considered:

$$S_X^{ss}(f) = 2S_X(f) \quad \text{if} \quad f > 0;$$
$$S_X^{ss}(0) = S_X(0).$$

Cross-power spectral density Having two signals $x(t)$ and $y(t)$, the cross-power spectral density is the limit, if existing,

$$S_{YX}(f) = \lim_{T \to +\infty} \frac{1}{T}Y_T^*(f) \cdot X_T(f) \qquad U_X U_Y\,Hz^{-1}.$$

---

[10]The attempt is done to achieve expressions which might more immediately translated in their correspondent discrete versions, suitable to be implemented in sampling systems, see Ch. 5.

[11]For a real $x(t)$, $X(f)$ is Hermitian: that is, $X^*(-f) = X(f)$, where X^* is the complex conjugate of X.

Unlike the (auto)power spectral density, the cross-power spectral density is a complex function, and $S_{YX}(f) = S^*_{XY}(f)$.

1.7.2   Impedance

Consider again the two-terminal element of Fig. 1.1 as a signal-processing system, having current $i(t)$ as its input and voltage $v(t)$ as its output.[12] Let us assume that the element is

linear: if $[i(t), v(t)]$ is a possible input-output signal pair for the element, then any $[ki(t), kv(t)]$ with real k is another possible couple. The equilibrium condition $[i(t) \equiv 0, v(t) \equiv 0]$ exists for all linear elements.

causal: effects on output occur *after* the input, so $i(\tau)$ has effect on $v(t)$ only if $\tau \leq t$;

time-invariant: the input-output relationship is stable over time, so if $[i(t), v(t)]$ is an input-output signal couple, $[i(t+T), v(t+T)]$ for an arbitrary T is another.

Given the properties above, the most general input-output relation can be written as

$$v(t) = \int_{\tau=-\infty}^{t} \zeta(t-\tau)\, i(\tau)\mathrm{d}\tau. \tag{1.16}$$

If we specify that $\zeta(\tau) \equiv 0$ for $\tau < 0$; or, equivalently, that $\zeta(\tau)$ must have the form $\zeta(\tau) = H(\tau) \cdot \xi(\tau)$, where $\xi(\tau)$ is a generic function and $H(\tau)$ is the *Heaviside function*

$$H(\tau) = \begin{cases} 0 & \text{if} \quad \tau < 0 \\ 1 & \text{if} \quad \tau \geq 0, \end{cases}$$

then Eq. (1.16) can be rewritten as a *convolution* integral:

$$v(t) = \int_{\tau=-\infty}^{+\infty} \zeta(t-\tau)\, i(\tau)\mathrm{d}\tau = \zeta(t) * i(t). \tag{1.17}$$

$\zeta(t)$ is called the *kernel* of the linear relation, or the *impulse response* of the element, since if the input is an unit current pulse $\delta(t)$ of $1\,\mathrm{A\,s}$ amplitude, then the corresponding voltage output waveform is $\zeta(t)$ (expressed in V). In SI, $[\zeta(t)] = \Omega\,\mathrm{s}^{-1}$.

Let us consider the amplitude spectra of $v(t)$, $i(t)$, and $\zeta(t)$:

$$V(f) = \mathcal{F}\,[v(t)] \qquad \mathrm{V\,Hz}^{-1},$$
$$I(f) = \mathcal{F}\,[i(t)] \qquad \mathrm{A\,Hz}^{-1},$$
$$Z(f) = \mathcal{F}\,[\zeta(t)] \qquad \Omega.$$

---

[12]Here only deterministic effects are considered; any noise is neglected.

Eq. (1.17) can be written as

$$V(f) = Z(f) I(f);$$

The definition of impedance can thus be extended beyond the sinusoidal steady state, to the most general voltage and current signals. The impedance function $Z(f)$ gives the relationship between the components at frequency f between signals $i(t)$ and $v(t)$.

1.7.3   An example: ideal elements

To give a few examples, let us derive again the impedance of ideal capacitance, resistance, and inductance:

capacitance is described by the relation $q(t) = Cv(t)$, or

$$v(t) = \int_{\tau=-\infty}^{t} \frac{1}{C} i(\tau)d\tau = \int_{\tau=-\infty}^{+\infty} \frac{1}{C} H(t-\tau) i(\tau)d\tau, \qquad (1.18)$$

and we can identify in Eq. (1.18) the term $\zeta(t) = \frac{1}{C}H(\tau)$, whose Fourier transform is

$$Z(f) = \mathcal{F}\left[\frac{1}{C}H(\tau)\right] = \frac{1}{2C}\delta(f) - i\frac{1}{2\pi f\, C}. \qquad (1.19)$$

The imaginary part of $Z(f)$ in (1.19) is the capacitive reactance

$$X(f) = -(2\pi f\, C)^{-1}.$$

The real term $\delta(f)/(2C)$ in Eq. (1.19) is there because the pure capacitor is a degenerate network (it has, in fact, infinite memory and thus not strictly linear, since admits many solutions $v(t) = K$ for $i(t) \equiv 0$). The term is conventionally omitted in the expression of capacitive impedance.[13]

resistance is described by (using the property $\int_\tau \delta(\tau)d\tau = 1$))

$$v(t) = Ri(t) = \int_{\tau=-\infty}^{\infty} R\,\delta(t-\tau)\, i(\tau)d\tau \rightarrow \zeta(t) = R\delta(t) \rightarrow Z(f) = R;$$

inductance is described by (using a property of the derivative $\delta'(t)$ of $\delta(t)$)

$$v(t) = L\frac{di(t)}{dt} = \int_{\tau=-\infty}^{\infty} L\,\delta'(t-\tau)\, i(\tau)d\tau \rightarrow \zeta(t) = L\delta'(t) \rightarrow Z(f) = L2\pi i f.$$
$$(1.20)$$

In Eq. (1.20) one can identify the inductive reactance $X(f) = 2\pi f\, L$.

---

[13]It can however be interpreted as the zero-frequency component of the response to a unitary current impulse (an unit charge) applied at $t = 0$. Suppose $C = 1\,\mathrm{F}$; then $v(t) = 0$ for $t < 0$ and $v(t) = 1\,\mathrm{V}$ for $t > 0$, and the average dc value over the whole time axis is $0.5\,\mathrm{V}$.

1.7.4   Linear response theorems

Within linear response theory, several general theorems can be demonstrated. In the following, a couple of them are introduced.

1.7.4.1   Kramers-Krönig relations

The dependence with frequency of real and imaginary parts of impedance $Z(\omega) = R(\omega) + \mathrm{j}X(\omega)$ are related by integral relations, called *Hilbert transformations* or *Kramers-Krönig relations* (Krönig, 1926; Kramers, 1927; Landau and Lifshitz, 1960; Bechhoefer, 2011):

$$R(\omega) = \frac{2}{\pi} P \int_0^\infty \frac{\omega' X(\omega')}{(\omega')^2 - \omega^2} \mathrm{d}\omega'; \qquad (1.21)$$

$$X(\omega) = -\frac{2\omega}{\pi} P \int_0^\infty \frac{R(\omega')}{(\omega')^2 - \omega^2} \mathrm{d}\omega', \qquad (1.22)$$

where P denotes the *Cauchy principal value* of the integral.

Relations (1.21) and (1.22) imply that the frequency dispersion of the primary parameter is tightly related to the magnitude of the secondary parameter. In principle, complete information about one parameter is sufficient to compute the other one; an example of computation on simulated data is shown in Fig. 1.9.

A direct calculation with Eq. (1.21) and (1.22) is not practical, because of diverging integrals; however, a calculation using direct and inverse Fourier transforms (Peterson and Knight, 1973) is easy to implement.

In practice, because of the necessarily limited accuracy and frequency domain of a real measurement, the application of Eq. (1.21) and (1.22) require careful hypotheses on the behavior of the measurand. Kramers-Krönig relations are widely employed in electrochemical impedance spectroscopy (Barsoukov and Macdonald, 2005, Ch. 3), but have until now found little application in impedance metrology (Zimmerman et al., 2006).

1.7.4.2   Fluctuation-dissipation theorem, and Johnson-Nyquist noise

Consider the two-terminal element of Fig. 1.10, having impedance $Z(f)$ (admittance $Y(f)$), in equilibrium (that is, not connected to external sources, or having stored energy in their reactance) with a thermal reservoir at the thermodynamic temperature Θ. The element of Fig. 1.10 develops a noise voltage (current) $v(t)$ ($i(t)$), called *Johnson-Nyquist noise* (Johnson, 1928; Nyquist, 1928).

The *fluctuation-dissipation theorem* (Callen and Welton, 1951) states that the noise generation at thermal equilibrium (fluctuation) has the very same physical origin of the dissipation occurring when the element is energized by an external source, related to the equivalent series resistance $R(f)$ (equivalent

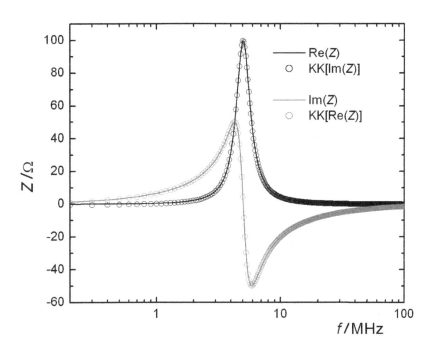

Figure 1.9
An example of application of Eq. (1.21) and (1.22). Impedance spectrum $Z(f)$
of a RLC parallel circuit (R =100 Ω, L =1 μH, C =1 nF) has been simulated
and plotted (black line: real part Re Z; gray line: imaginary part Im Z). Eq.
(1.21) and (1.22) have been evaluated numerically in a wide frequency range
(100 kHz − 1 GHz, 10000 frequency points). With Eq. (1.21), Kramers-Krönig
(KK) estimates KK [Im Z] of Re Z, and KK [Re Z] of Im Z have been computed
and plotted (black and gray ○ symbols, respectively).

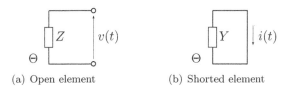

Figure 1.10
(a) Open element and (b) short-circuited element in equilibrium with a thermal bath at temperature Θ.

parallel conductance $G(f)$). The noise has a Gaussian amplitude distribution and its power spectral density $S_v(f)$ ($S_i(f)$), see Sec. 1.7.1, is given by

$$S_v(f) = 4k_B\,R(f)\,\Theta\,p(f) \qquad \text{V}^2\,\text{Hz}^{-1};$$
$$S_i(f) = 4k_B\,G(f)\,\Theta\,p(f) \qquad \text{A}^2\,\text{Hz}^{-1}, \qquad (1.23)$$

where k_B is the Boltzmann constant, and $p(f)$ is the Planck factor[14]

$$p(f) = \frac{hf}{k_B\Theta}\,\frac{1}{\exp\left(\frac{hf}{k_B\Theta}\right) - 1}, \qquad (1.25)$$

dependent on f, Θ, k_B and the Planck constant h (see Appendix C). Fig. 1.11 gives a plot of Eq. (1.25).

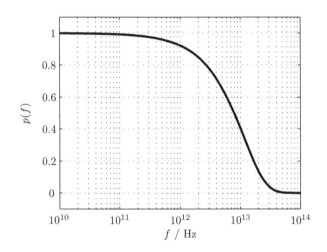

Figure 1.11
A plot of the Planck factor $p(f)$, Eq. (1.25), for $\Theta = 300\,\text{K}$.

At $f > h^{-1}\,k_B\,\Theta$ (at room temperature, in the THz frequency range), *quantum cutoff* occurs.

If f is decades below quantum cutoff, $p(f) \approx 1$ (for example, at room temperature $|p(f) - 1| < 10^{-6}$ up to f in the MHz frequency range), and Eq.

---

[14]Planck factor (1.25) can be written in the modified form

$$p(f) = \frac{hf}{k_B\Theta}\left[\frac{1}{\exp\left(\frac{hf}{k_B\Theta}\right) - 1} + \frac{1}{2}\right] = \frac{hf}{k_B\Theta}\coth\left(\frac{hf}{2k_B\Theta}\right); \qquad (1.24)$$

the term $\frac{1}{2}$ in (1.24) is related to *zero-point* energy, see, e.g., Callen and Welton (1951). Whether the zero-point energy term give experimentally observable noise is matter of scientific debate (Abbott et al., 1996).

(1.23) can be approximated by its classical white-noise expression:

$$S_v(f) = 4k_B\, R(f)\,\Theta, \qquad S_i(f) = 4k_B\, G(f)\,\Theta. \qquad (1.26)$$

An extension of the fluctuation-dissipation theorem to multi-terminal networks also exists (Twiss, 1955; Büttiker, 1990).

Johnson noise gives a fundamental limit to the sensitivity of all electrical measurements, including of course impedance measurements.

1.7.5   Non-equilibrium noise

Energizing an impedance during measurement can generate further non-equilibrium noise, called *excess noise*. Excess noise depends on the particular impedance considered (that is, not only on its value but its particular construction), and on the energizing conditions. We only quickly mention some non-equilibrium noises having special names; they were discovered before Johnson noise.

Shot noise (Schottky, 1918, 1926), current noise caused by localization of electric charge in the impedance considered, usually because of semiconductor or tunnel junctions. In the dc steady-state approximation (constant current I), it's a non-Gaussian white noise having power spectral density $S_i(f) = 2q\, I\, F$, where q is the current carrier elementary charge (usually the electron charge e), and F is the *Fano factor*, an adimensional quantity having values in the $0 \ldots 2$ range.

Flicker, or $1/f$, noise (Johnson, 1922; Schottky, 1926), a ubiquitous low-frequency noise having the general power spectral density form $S(f) \sim f^{-\alpha}$, where $\alpha \approx 1$. In impedance measurement circuits, all devices, including electronic components, such as reference sources (Witt et al., 1995; Witt, 1997) and detectors, generate flicker noise. The impedance under measurement can also generate flicker noise (Voss and Clarke, 1976): see Fig. 1.12.

Barkhausen noise (Barkhausen, 1919), a voltage noise occurring in devices, including ferromagnetic materials (such as iron-core inductors, transformers), when the applied current is smoothly changed (e.g., during sinusoidal steady state typical of impedance measurements). It is caused by steps in the magnetic flux-field characteristic of the ferromagnetic material, in turn due to sudden reversals of magnetic domains. Barkhausen noise is a low-frequency noise, and its typical power spectral density is negligible above 1 kHz. For a recent review, see Durin and Zapperi (2005).

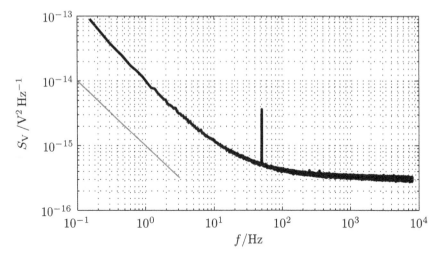

Figure 1.12
Noise of a $20\,\mathrm{k\Omega}$ carbon resistor at room temperature, when a low-noise dc current excitation $(160\,\mathrm{\mu A})$ is applied. Low-frequency spectrum is dominated by flicker noise (the gray line shows $1/f$ dependence for comparison); at higher frequencies, thermal noise of the resistor dominates. The peak at $50\,\mathrm{Hz}$ is interference from mains.

1.8 Impedance as an electromagnetic quantity

In the previous sections, we treated impedance as a purely phenomenological quantity relating voltage and current in a network. Here we link impedance to electromagnetism, for the case of a one-port linear passive network in the sinusoidal steady state.[15] Refer to Appendix B for a derivation of Maxwell equations and Poynting theorem for harmonic electromagnetic fields.

1.8.1 One-port impedance

Let us consider the passive, linear, one-port electromagnetic system of Fig. 1.13. In the sinusoidal steady state at angular frequency ω, its coaxial port is fed with voltage $V_i(\omega)$ and current $I_i(\omega)$,[16] so its impedance is $Z(\omega) = V_i(\omega)/I_i(\omega) = R(\omega) + jX(\omega)$.

The electromagnetic power P_i, entering the system through the port sur-

---

[15]The following treatment follows that of Jackson (1975, Ch. 6.10), rewritten in SI units.
[16]It is implicitly assumed that the electromagnetic field pattern at the port is the TEM (Transverse ElectroMagnetic) mode.

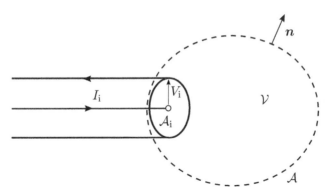

Figure 1.13
A one-port linear, passive electromagnetic system, delimited by volume $\mathcal{V}$ and
area $\mathcal{A}$ having normal n. Its input port, of area $\mathcal{A}_i$, is fed by voltage V_i and
current I_i.

face $\mathcal{A}_i$, is

$$P_i = \frac{1}{2} I_i^* V_i = - \int_{\mathcal{A}_i} S(r) \cdot n(r) \, d\mathcal{A}_i,$$

where we put the minus sign because the input power has direction opposite
to the surface normal $n(r)$.

The Poynting theorem, Eq. (B.4), can be written as

$$\frac{1}{2} I_i^* V_i = \frac{1}{2} |I_i^2| \, (R + jX)$$

$$= \frac{1}{2} \int_{\mathcal{V}} J^*(r) \cdot E(r) d\mathcal{V} + 2j\omega \int_{\mathcal{V}} (w_m - w_e) \, d\mathcal{V} + \int_{\mathcal{A} - \mathcal{A}_i} S(r) \cdot n(r) \, d\mathcal{A}.$$

It can be shown that the integral $\int_{\mathcal{A}-\mathcal{A}_i} S(r) \cdot n(r) \, d\mathcal{A}$ is real and represents
the *radiation loss* through area $\mathcal{A}-\mathcal{A}_i$. By separating real and imaginary parts,

$$R = \frac{1}{|I^2|} \left(\operatorname{Re} \left[\int_{\mathcal{V}} J^* \cdot E \, d\mathcal{V} \right] + 4\omega \operatorname{Im} \left[\int_{\mathcal{V}} (w_e - w_m) \, d\mathcal{V} \right] + 2 \int_{\mathcal{A}-\mathcal{A}_i} S(r) \cdot n(r) \, d\mathcal{A} \right),$$

$$X = \frac{1}{|I^2|} \left(\operatorname{Im} \left[\int_{\mathcal{V}} J^* \cdot E \, d\mathcal{V} \right] + 4\omega \operatorname{Re} \left[\int_{\mathcal{V}} (w_m - w_e) \, d\mathcal{V} \right] \right). \tag{1.27}$$

The terms in Eq. (1.27) have the following meaning:

- $\operatorname{Re} \left[\int_{\mathcal{V}} J^* \cdot E \, d\mathcal{V} \right]$ is the power dissipated by *Joule effect*;

- $4\omega \operatorname{Im} \left[\int_{\mathcal{V}} (w_e - w_m) \, d\mathcal{V} \right]$ is the *loss* in dielectric and magnetic materials
  in $\mathcal{V}$ because of internal dissipation caused by the varying fields;

- $2 \int_{\mathcal{A}-\mathcal{A}_i} S(r) \cdot n(r) \, d\mathcal{A}$ is the power radiated by the electromagnetic field,
  and the associated resistance term is the *radiation resistance*;

- Im $\left[\int_{\mathcal{V}} \boldsymbol{J}^* \cdot \boldsymbol{E} \, \mathrm{d}\mathcal{V}\right]$ is the energy stored because of a delay of the current density in following the driving field, because charge carriers have inertia; the associated reactance is the *kinetic reactance*;

- $4\omega \, \mathrm{Re}\left[\int_{\mathcal{V}} (w_{\mathrm{m}} - w_{\mathrm{e}}) \, \mathrm{d}\mathcal{V}\right]$ is the energy stored in the electromagnetic field; the associated reactance terms are the *inductive* and *capacitive reactance*, having opposite signs.

1.9   The graphical representation of electrical impedance

Electrical impedance $Z(f)$ (admittance $Y(f)$), a complex quantity, can be graphically represented in several ways. Refer to Figs. 1.14–1.19.

real and imag plot, Fig. 1.14, is the plot of the resistance $R(f) = \mathrm{Re} \, Z(f)$ and reactance $X(f) = \mathrm{Im} \, Z(f)$ (or conductance $G(f) = \mathrm{Re} \, Y(f)$ and susceptance $B(f) = \mathrm{Im} \, Y(f)$) versus frequency.

Bode plot, Fig. 1.15, is the graph of the complex quantity $Z(f)$ (or $Y(f)$) with a log-frequency axis. It is usually a combination of a Bode magnitude plot of and a Bode phase plot. If $Z(f)$ (or $Y(f)$) is a rational function with real poles and zeros, then the Bode magnitude plot can be approximated with lines having slope f^k ($k = 0, \pm 1, \pm 2, \ldots$), which appear as straight lines in a log-log plot. If a simple equivalent circuit of a passive component is available, rules exist to draw approximated Bode magnitude and phase plots by hand.

Nyquist plot, also called *Cole-Cole plot* (Cole and Cole, 1941, 1942), Fig. 1.16, is a parametric graph of $Z(f) = R(f) + \mathrm{j}X(f)$ (or $Y(f) = G(f) + \mathrm{j}B(f)$) on the $R - X$ ($G - B$) complex plane, typically with equal axis scaling. The resulting graph is a two-dimensional curve, having f as implicit parameter. If $Z(f)$ is a rational function having poles that are well separated in frequency, the Nyquist plot is composed of semicircles (for first-order poles, as in RL or RC series or parallel) and full circles (for second-order poles, as in LC series or parallel) jointed together, see also Fig. 1.17.

Smith chart, a plot of the *normalized impedance* $z(f) = Z(f)/Z_0$, where Z_0 is the reference impedance of choice (typically, $50 \, \Omega$). Smith chart is actually an orthogonal, equal axis Cartesian plot of the value $\Gamma(f)$, the *reflection coefficient*, defined from $z(f)$ by the transformation $z \mapsto \Gamma$

$$\Gamma(f) = \frac{z(f) - 1}{z(f) + 1}. \tag{1.28}$$

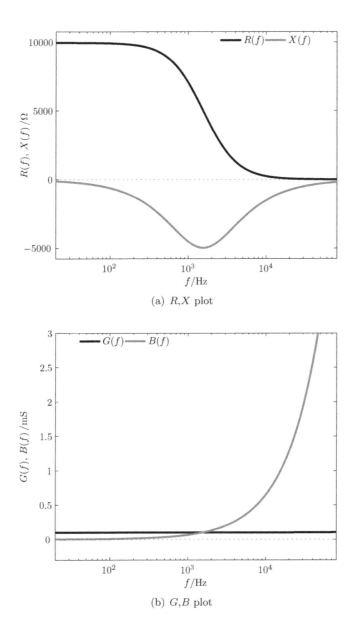

(a) *R,X* plot

(b) *G,B* plot

Figure 1.14
Different representations of impedance. Measurement of a 2T impedance (a *RC* parallel of electronic components: $R = 10\,\mathrm{k\Omega}$, $C = 10\,\mathrm{nF}$), over a wide frequency range, with an Agilent E4980A RLC bridge. Plot of (a) $R(f)$ and $X(f)$; (b) $G(f)$ and $B(f)$ versus frequency f.

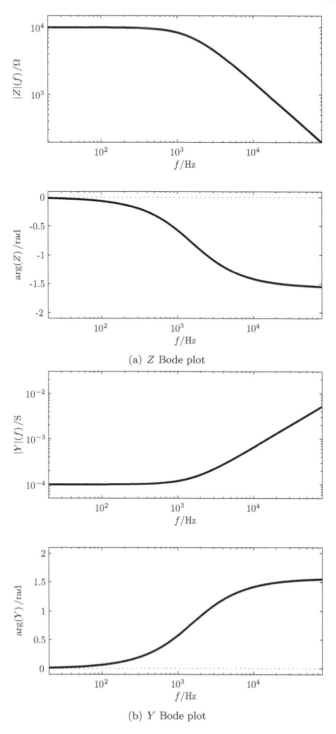

(a) Z Bode plot

(b) Y Bode plot

Figure 1.15
Same data of Fig. 1.14, Bode plots.

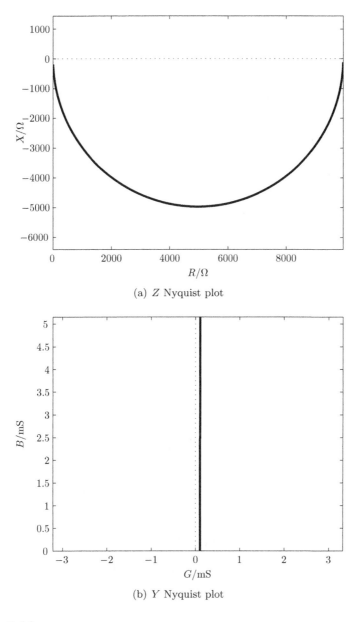

(a) *Z* Nyquist plot

(b) *Y* Nyquist plot

Figure 1.16
Same data of Fig. 1.14, Nyquist plots.

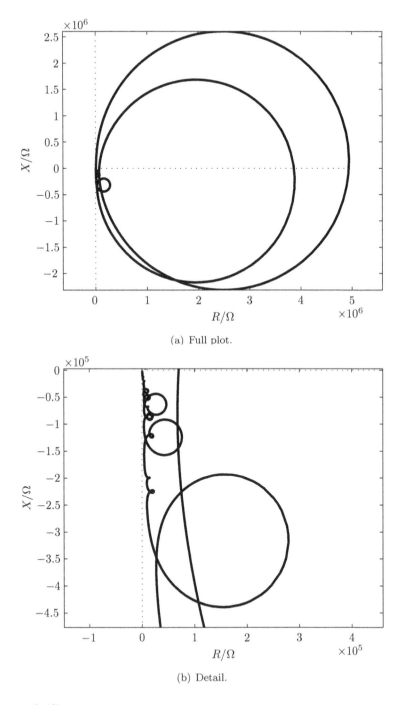

(a) Full plot.

(b) Detail.

Figure 1.17
Nyquist Z plot of impedance measured on a high-valued inductor (10 H air-core toroidal inductor General Radio mod. 1482-T, measured with Agilent E4980A RLC bridge in the 20 Hz − 2 MHz frequency range). Each LC resonance of the winding gives a full circle in the graph.

In the Smith chart, the usual Cartesian coordinate system is replaced with the $z \mapsto \Gamma$ transformation of the original z-plane coordinate system.

Eq. (1.28) is a particular *Möbius transformation*, mapping the extended complex plane (thus including the point ∞) into itself. Möbius transformations are *conformal* (that is, angle-preserving) and map generalized circles (ordinary circles and straight lines) into generalized circles.

The $z \mapsto \Gamma$ transformation maps the whole right-half complex plane (that is, $\text{Re}\, z \geq 0$) on a unity-radius circle centered on the origin. In particular, $z = 1 + \text{j}0$ (that is, $Z = Z_0$) maps to the origin $\Gamma = 0 + \text{j}0$. Real impedances ($\text{Im}\, z = 0$) are located on the horizontal diameter, starting with zero impedance $z = 0 + \text{j}0$ on the left and ending with $z = +\infty + \text{j}0$ on the right. Inductive impedances ($\text{Im}\, z \geq 0$) are located on the upper semicircle; capacitive impedances ($\text{Im}\, z \leq 0$) on the lower one.

If $z(f)$ has constant resistance or reactance, its plot is an arc of circumference; the same if $z^{-1}(f)$ has constant conductance or susceptance; the Smith chart coordinate system is in fact given by the set of these curves.

The Smith chart is a handy graphical calculation tool for a number of problems and is widely employed in radio frequency applications.

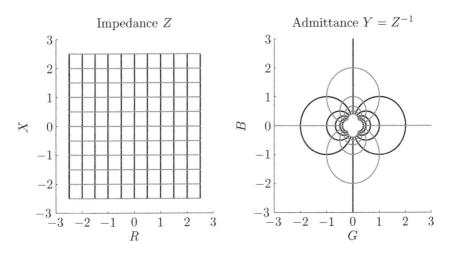

Figure 1.18
The inversion transformation $Y = Z^{-1}$ applied to a finite square grid in the Z plane.

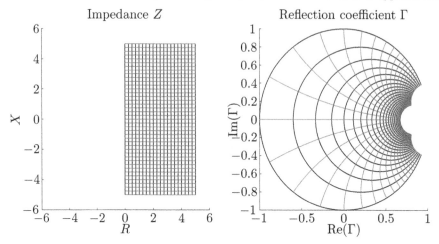

Figure 1.19

The transformation $\Gamma = \dfrac{Z/Z_0 - 1}{Z/Z_0 + 1}$, where Z_0 is the characteristic impedance (here $Z_0 = 1 + i0$), applied to a finite square grid in the $R \geq 0$ portion of the Z plane. The resulting transformed grid is the *Smith chart*.

1.10   Density quantities

The impedance of an artifact is dependent on the electromagnetic properties of the material composing it. By appropriate *fixtures* and a measurement model, see Sec. 6.1, electromagnetic density quantities can be derived from impedance and geometrical measurements, see Sec. 6.1.

Volume- or surface-density quantities can be considered; *scalar* quantities for isotropic materials become *tensor* quantities for anisotropic ones (the tensor rank is dependent on the dimensionality, 3 for volume quantities, 2 for surface quantities).

Volume density quantities related to impedance are:

conductivity σ (S m^{-1}). It relates current density $\boldsymbol{J}$ to electric field $\boldsymbol{E}$ via Ohm's law $\boldsymbol{J} = \sigma \cdot \boldsymbol{E}$.

Typical values range from 60 MS m^{-1} of best conducting metals (copper and silver) to 5.5 µS m^{-1} of ultra-pure water (for best insulators, *resistivity* is instead typically quoted).

The conductivity of metal alloys can also be expressed with the *International Annealed Copper Standard* (IACS) scale, as a percentage. 100 % IACS corresponds to a conductivity of 58 MS m^{-1}, the conductivity of annealed copper at 20 °C.

resistivity $\rho = \sigma^{-1}$ (Ω m). Typical values range from $16\,\text{n}\Omega\,\text{m}$ of copper and silver, to $10\,\text{Z}\Omega$ mdash$1000\,\text{Z}\Omega$ m of best insulators (Teflon).

permittivity, or dielectric constant, ϵ or sometimes κ ($\text{F}\,\text{m}^{-1}$). It relates the dielectric induction field D with the electric field E via the relation $\boldsymbol{D} = \epsilon \cdot \boldsymbol{E}$; normally a scalar quantity, it becomes a tensor in anisotropic materials.

A related quantity is the *relative permittivity* $\epsilon_r = \epsilon/\epsilon_0$, where ϵ_0 is the *vacuum permittivity*, or *electric constant*, see App. C. Typical relative permittivity values for insulators range from 1 (vacuum) to 10^3 for high-κ dielectrics.

permeability μ ($\text{H}\,\text{m}^{-1}$). It relates the magnetic induction field B with the magnetic field H via the relation $\boldsymbol{B} = \mu \cdot \boldsymbol{H}$; normally a scalar quantity, it becomes a tensor in anisotropic materials.

A related quantity is the *relative permeability* $\mu_r = \mu/\mu_0$, where μ_0 is the *vacuum permeability*, or *magnetic constant* see App. C. For many materials, μ_r values cluster around the vacuum permeability $\mu_r = 1$ (0.999 834 for bismuth, a *diamagnet*, to 1.000 265 for platinum, a *paramagnet*). *Superconductors* have $\mu_r = 0$ (superdiamagnetism, or Meissner effect), and *soft ferromagnets* can have μ_r ranging from 100 (magnetic steel) to 100 000 (mu-metal). In ferromagnets, however, the $B - H$ relation is in general no more linear and μ_r is strongly dependent on H and the past magnetic history of the material; the very meaning of permeability and its measurement must therefore be carefully considered. See Sec. 6.1.2 for a more extended discussion.

Surface density quantities are relevant when the material of interest is "two-dimensional" (has one dimension much smaller than the other two and of constant magnitude, as in uniform-thickness thin sheets and films) or when the electrical properties of the surface of the three-dimensional material are significantly different than the bulk, because of chemical or physical contamination of the surface (for example, oxidation or humidity absorption).

Surface density quantities related to impedance are

surface conductivity σ_S (S).

surface resistivity $\rho_S = \sigma_S^{-1}$ (Ω);

All density quantities, in the sinusoidal steady state, can be considered as complex quantities having both real and imaginary parts; for example, conductivity $\sigma(\omega)$ can be expressed as $\sigma(\omega) = \sigma'(\omega) + j\sigma''(\omega)$. Someone talks of *impeditivity* $\rho(\omega) = \rho'(\omega) + j\rho''(\omega)$, and $\rho''(\omega)$; in the same context, $\rho''(\omega)$ is called *reactivity*.

1.10.1   Penetration depth and skin effect

The *penetration depth* is the characteristic length of attenuation of a magnetic field impinging on a conductor due to the onset of *eddy currents*; it is directly related to σ and μ of the conducting material. For a semi-infinite slab of material, and an applied ac magnetic field at frequency f parallel to the surface, the penetration depth is $\delta = (\pi f \mu \sigma)^{-\frac{1}{2}}$. The magnetic field can be generated by an electric current in the conductor itself. The resulting phenomenon is called *skin effect*, a redistribution of current density in the conductor. Circular wires having diameter $D \gg \delta$ have a resistance to ac currents, which is approximately the dc resistance of a hollow tube with wall thickness δ.

1.11   Frequency ranges

Concepts defined in the present chapter, in particular those related to network theory, are valid for any working frequency. However, here and, more often, in the following chapters, the feasibility of application of a particular impedance representation, definition (Ch. 2), or measurement method (Ch. 4 and 5), and the availability of suitable instruments, devices, and impedance standards (Ch. 3 and 8) can depend on the frequency range of interest.

Although a precise classification of frequency bands in the radio spectrum is widely employed in communication, it is too detailed for impedance measurement. Throughout the book, the following informal terms will be employed:

direct current (dc) for an excitation (either voltage or current) that is kept at the same level for measuring times much longer than all time constants of the measurement setup employed. Often, in dc measurements, the signal polarity is periodically reversed at very low frequency (typically $\ll 1\,\mathrm{Hz}$);

power, or **mains**, frequency: 50 Hz or 60 Hz, or any frequency near these values;

low, or **audio**, frequency: between 20 Hz and 100 kHz;

high, or **radio (RF)**, frequencies: between 100 kHz and 1 GHz;

microwave (MW) frequencies: from 1 GHz to 100 GHz and beyond.

The distinction between frequency ranges is increasingly blurred by the availability of commercial instruments that span broader and broader bandwidths. The problem of achieving accurate measurement methods and proper traceability in the frequency range from 100 kHz to 100 MHz is a subject of present impedance metrology research often called the *LF-RF gap* problem (Callegaro, 2009).

2

Impedance definitions

CONTENTS

In the measurement of electrical impedance, the *definition* of the measurand plays an essential role. In metrology, the definition is the set of rules to identify in a unique way the *measurand*: "...the kind of quantity, description of the state of the phenomenon, body, or substance carrying the quantity, including any relevant component..." (VIM, 2.3). The purpose of impedance definition is to define a geometrical closed surface and a set of electrical boundary conditions, in the attempt to exclude every effect of connections, cables, the meter, and electromagnetic properties of the environment, so that changes in the latter have no effect on the measurement result (VIM, 2.9).

Here we consider as subject of measurement physical devices, electrically accessible by the outside world through a number of conductors; the geometrical surface of the definition intersects the conductors at well-defined sections a, b, c, ... In impedance standards, the geometrical surface can often be identified by the case, which is often also a *shield*, and the sections are defined by connectors mounted on the case. We restrict the discussion to devices that can be modeled as linear passive electrical networks, for which (at given frequency *f*) the concept of transimpedance has been defined in Sec. 1.6.1. We adopt as definition of impedance Z the quantity $Z = V_{ab}/I_c$, where voltage V_{ab} is measured between sections a and b of two conductors, and current I flowing through the section c of a third conductor.

Impedance definition must include rules to identify a, b, c, and a number of electromagnetic conditions that must be satisfied at the geometrical boundary surface.

Typically, when dealing with components or devices, the operator has some choice on the more appropriate impedance definition, and on the location of a, b, c. In impedance standards, instead, the definition and location of a, b, c is the direct consequence of the numbering, physical location, and type of connections available, which are thus accordingly labeled.

Impedance metrology distinguishes between *n-terminal* and *n-terminal-pair* definitions.

2.1   *n*-terminal definitions

In "*n*-terminal" definitions of impedance, points a and b are geometrically separated and located on the length of two conductors, called *terminals*; section c is a given cross section of a terminal (either the same terminal, where a or b are located, or a third one). *Two-terminal, three-terminal, four-terminal*, and *five-terminal* definitions are commonly employed in dc resistance and impedance metrology at mains power frequency. Typically, measurement setups based on *n*-terminal impedance definitions employ binding posts as connectors, and unshielded electrical cables. Simple electromagnetic boundary conditions apply.

2.1.1   Two-terminal

In two-terminal definition (2T), Fig. 2.1(a), the impedance Z interfaces with the meter by two fully equivalent conductors. The conductors carry the current I, and voltage V is measured between the conductors; the current on either one of them, see Fig. 2.1(b). No electromagnetic boundary conditions apply.

All stray parameters among the impedance body, the enivronment, and the wiring have an influence on the measurement. Fig. 2.1(b) shows, as lumped parameters, the capacitance between impedance body and each wiring conductor versus "ground" (the environment average potential), the resistance, self- and mutual-inductance, and cross-capacitance of the wiring. All these parameters are dependent on the wiring properties, its geometrical positioning, and the general geometry of the environment (conducting surfaces, including the operator, and ferromagnetic materials have the most important effects). In general, it is difficult to measure or model such parameters, or even to reproduce them in two different measurement setups, except when rigid connections

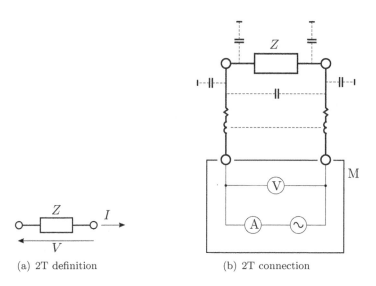

(a) 2T definition

(b) 2T connection

Figure 2.1
(a) Two-terminal (2T) definition of impedance. (b) The connection of a 2T impedance standard to a meter M.

Figure 2.2
An example of a 2T impedance, a Vishay mod. S9570 $10\,\text{M}\Omega$ resistor. The metal case is left unconnected.

and careful electrostatic shielding of the experiment is involved (Hanke et al., 2002).

An example of a two-terminal impedance standard is shown in Fig. 2.2.

2.1.2   Three-terminal

A three-terminal impedance (3T), Fig. 2.3(a), is enclosed in a conducting surface: the *shield*, or *guard*, accessible with an additional terminal G. An example of a 3T impedance is shown in Fig. 2.4.

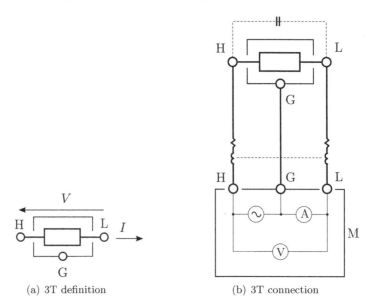

(a) 3T definition   (b) 3T connection

Figure 2.3
(a) Three-terminal (3T) definition of impedance. (b) The connection of a 3T impedance standard to a meter M.

The purpose of the shield is to define (thus, provide independence with respect to the particular environment) the stray capacitance and loss currents from the impedance body and the environment. In order to achieve this, the terminals are no more equivalent (as they were in the two-terminal definition) but are marked: typically, as *high* H and *low* L.

The electromagnetic boundary condition $V_{LG} = 0$ apply:[1] the current is measured at the L terminal. Typically, G is at a potential similar to the surrounding environment.[2]

---

[1]This does not mean that L and G are shorted together. The meter M has to reach the definition condition by construction or with the injection of appropriate voltages or currents in the circuit mesh.

[2]In applications concerning dc high resistance – or low current – measurement setups, G

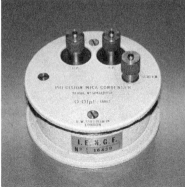

(a) Sullivan 10 nF capacitor

(b) Detail of a General Radio mod. 1433-W resistance box

Figure 2.4
Examples of 3T impedance.

The connection of a three-terminal impedance to a meter M is shown in Fig. 2.3(b); the lumped wire impedances show that three-terminal definition is still sensitive to all resistances, self- and mutual inductances of the wiring.

2.1.3 Four-terminal

The aim of the *four-terminal* (4T) definition of resistance and impedance is to exclude the series parameters (resistance and self-inductances) of all connections from the measurement. It was first introduced by William Thomson, Lord Kelvin, in 1860 and is largely employed in low-value dc resistance measurements (Thomson, 1860).

In a 4T impedance, two equivalent *current* terminals (C) are connected to the current generator of the meter; voltage drop due to the impedance is measured between two equivalent *voltage* or *potential* terminals (P). The 4T definition condition is $I_P = 0$; i.e., no current must be diverted by the voltage measurement circuitry.

As Fig. 2.5(b) shows, for what concerns stray capacitances between connections, and among connections and the environment, and the effect of mutual inductances (of great relevance when dealing with low-value impedances measured at high current values), the 4T definition faces problems similar to the 2T definition. An example of 4T impedance is shown in Fig. 2.6.

---

can be connected to a guard generator, driven to achieve $V_{LG} = 0$ condition. In such cases, often the guard is further shielded with another screen at low potential, and triaxial connections and cables are employed. Triaxial or multi-axial connections in impedance metrology occur rarely (Cabiati and D'Emilio, 1975; Cabiati and D'Elia, 2000, 2002; Ricketts et al., 2003), and a general theory of impedance metrology with triaxial connections has not yet been developed.

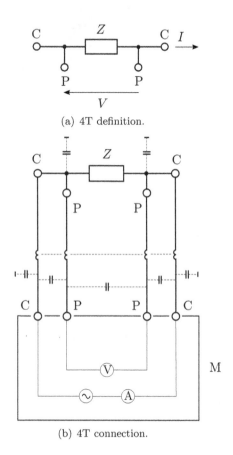

(a) 4T definition.

(b) 4T connection.

Figure 2.5
(a) Four-terminal (4T) definition of impedance. (b) The connection of a 4T
impedance standard to a meter M.

Figure 2.6
An example of 4T impedance standard, Tinsley mod. 5685A 25 Ω ac/dc resistor.

2.1.4 Five-terminal

Comparing Fig. 2.7(a) with Figs. 2.3(a) and 2.5(a), it is easy to see that the idea behind *five-terminal* definition is to merge 3T and 4T definitions to take adavantage of the benefits of both. As in 4T definition, current and voltage terminals are separated, but in 5T definition they are no more equivalent. In Fig. 2.7(a) current terminals are marked HC and LC, voltage terminals HP and LP; as in 3T definition, the impedance is enclosed in a shield connected to a guard terminal G, and the definition condition is $V_{\text{LP-G}} = 0$ applies.

The definition is commonly employed when dealing with synthesized impedance standards (Sec. 8.4), such as electromagnetic capacitors (Sec. 8.4.1) and capacitance calibrators (Sec. 8.4.3).

An example of a five-terminal standard is shown in Fig. 2.8.

2.1.5 Problems with *n*-terminal definitions

In terminal definitions, V_{ab} is assumed to be independent of the measurement geometry (i.e., independent of the integration path of the electric field). Such electrostatic approximation limits the applicability of terminal definitions to low frequency measurements.

Consider for example 4T impedance definition (the same argument is valid for all *n*-terminal definitions). The impedance voltage V occurs between potential terminals P placed at some distance, and connected with conductors having an arbitrary shape to meter M. The resulting voltage mesh has a contour Γ enclosing a surface S.

Any other circuit mesh k that carries current I_k (in the 4T example the mesh carrying the measuring current I), and that develops a magnetic field of which a fraction is linked to S, has a coupling with the voltage mesh, which

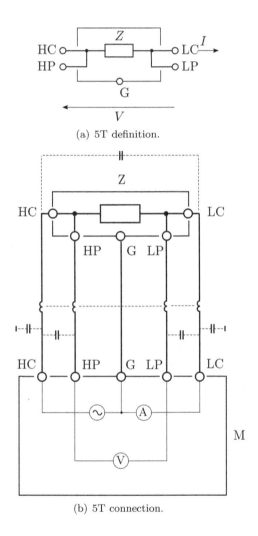

(a) 5T definition.

(b) 5T connection.

Figure 2.7
(a) Five-terminal (5T) definition of impedance. (b) The connection of a 5T impedance standard to a meter M.

Figure 2.8
An example of a five-terminal (5T) impedance standard, Fluke mod. 742A
$10 \, \mathrm{k\Omega}$ resistor.

can be modeled by a mutual inductance M_k. In the sinuosoidal steady state
at angular frequency ω, the voltage V_M measured by M is

$$V_\mathrm{M} = V + \mathrm{j} \sum \omega M_k I_k. \tag{2.1}$$

Eq. (2.1) shows that the measured voltage V_M is different from V. Such
effect can be minimized by choosing a geometry that minimizes S (e.g., by
twisting voltage leads). However, n-terminal definitions become increasingly
inadequate with increasing ω.

2.2 n-terminal pair definitions

2.2.1 Coaxial pairs

In n-terminal-pair, or n-port, impedance definition, the impedance is shielded,
and all n terminations are coaxial pairs, or *ports*. Typically, each coaxial pair
corresponds to a coaxial connector mounted on the shield (the outer conductor
of the coaxial connector is electrically connected to the shield). n-terminal-
pair impedances are defined assuming *coaxiality*; that is, for each port, its
inner and outer conductor carry the same current amplitude, but in opposite
directions. Port voltages are defined as the potential difference between the
inner and outer conductor, at a given cross section (typically the reference
plane of the coaxial connector).

n-terminal-pair impedances are connected to the measuring circuit with n
coaxial leads.

The conducting surface given by the impedance shield and outer conductors of coaxial leads is electrically continuous; such network acts as an electrostatic shield surrounding the circuit made by the impedance and inner conductors in the leads. Further, because of coaxiality, the magnetic field generated by current flowing is, to a first approximation, confined within the coaxial leads (see Sec. 3.5.2). Therefore, mutual inductances between leads are negligible, and the main problem of n-terminal definitions, mutual inductances among connections (Sec. 2.1.5) can be solved.

n-terminal-pair definitions often require a more complex measurement setup than n-terminal definitions.

2.2.2 One-terminal pair

In *one terminal-pair*, or *one-port* (1P) impedances, see Fig. 2.9(a), both voltage and current are defined on the same network port. Impedance is defined as $Z_{1P} \equiv V/I$. No defining conditions apply.

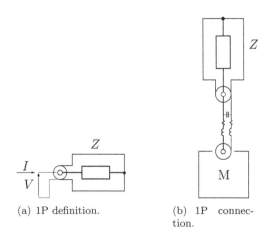

(a) 1P definition.

(b) 1P connection.

Figure 2.9
(a) One-terminal pair (1P) definition of impedance. (b) The connection of a 1P impedance standard to a meter M.

1P definition is widely employed in radio-frequency and microwave metrology, and has a direct physical meaning, see Sec. 1.8. The accuracy in the impedance definition is limited by the identification of the reference plane of the coaxial port permitted by the connector stray parameters and repeatability; Fig. 2.9(b) (right) shows that the meter M measures, in addition to Z, the resistance, inductance, and capacitance of the connecting coaxial line. Therefore, quality of the connectors employed in 1P measurements is of outmost relevance.

An example of 1P standard is shown in Fig. 2.10. Other examples are shown in Fig. 8.37.

Figure 2.10
Example of an 1P impedance, a General Radio 1406-D 100 pF gas-dielectric capacitor.

2.2.3 Two-terminal pair

In *two terminal-pair*, or *two-port* (2P) impedance definition, Fig. 2.11(a), the impedance standard is provided with two separate ports, labeled H and L.

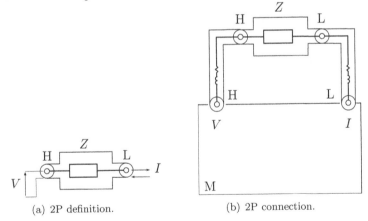

(a) 2P definition.　　　　　(b) 2P connection.

Figure 2.11
(a) Two-terminal pair (2P) definition of impedance. (b) The connection of a 2P impedance standard to a meter M.

Voltage is measured at port H, current at port L. Impedance is defined as

$$Z_{2P} = \frac{V_H}{I_L}, \quad \text{with the defining condition} \quad V_L \equiv 0.$$

Two terminal-pair definition is widely employed in low-frequency impedance

Figure 2.12
Example of a 2P impedance, a General Radio 1403-D 100 pF gas-dielectric capacitor.

metrology for high- and medium-value impedances, and find application also in high-resistance dc measurements. Fig. 2.11(b) shows that series stray parameters of the connections influence the measurement, which is instead insensitive to parallel parameters (stray capacitances); coaxiality eliminates mutual inductances between H and L connections.

An example of two terminal-pair standard is shown in Fig. 2.12.

2.2.4   Four-terminal coaxial

Four terminal coaxial (4C) impedance definition, Fig. 2.13(a), is a coaxial version of the 4T definition (Sec. 2.1.3).

The coaxiality permits to null mutual inductances between voltage and current connections. Current is measured at port I, voltage at port V; impedance is defined as

$$Z_{4C} = \frac{V_V}{I_I}, \quad \text{with the defining condition} \quad I_V \equiv 0.$$

4C definition is employed for low-value impedances, typically resistive shunts and thermometers, see Fig. 2.14.

2.2.5   Four-terminal pair

Four terminal-pair impedance definition, Fig. 2.15(a), is the most complete definition now in use in electrical metrology, combining the benefits of both

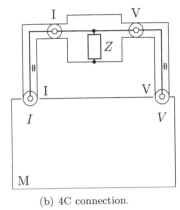

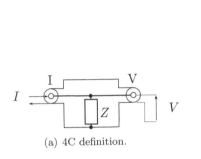

(a) 4C definition.   (b) 4C connection.

Figure 2.13
(a) Four-terminal coaxial (4C) definition of impedance. (b) The connection of a 4C impedance standard to a meter M.

2P and 2C definitions. It was introduced by Cutkosky (1964). A 4P standard is provided with four ports, in the following labeled[3] as HC, HP, LP, LC.

The impedance is energized at port HC by a generator; electrical parameters at HC are not measured. Voltage V_{HP} at port HP and current I_{LC} at port LC are measured and enter the definition.

4P impedance is defined as

$$Z_{4P} = \frac{V_{HP}}{I_{LC}},$$

with the defining conditions
$$I_{HP} \equiv 0,$$
$$V_{LP} \equiv 0,$$
$$I_{LP} \equiv 0.$$

The meaning of such defining conditions can be understood better by looking at Fig. 2.16 (Kibble, 1999).

An example of a 4P standard is shown in Fig. 2.17.

The main drawback of 4P definition is the difficulty in implementing the fulfillment of the defining conditions in the metering circuitry. Several commercial elecronic impedance meters are provided with feedback circuits to achieve 4P definition. In metrology-grade transformer bridges, instead, the condition

---

[3]More typical labeling of 4P standards in the literature is the enumeration 1, 2, 3, 4. However, such enumeration is not standard: compare older papers (Cutkosky, 1964; Kibble and Rayner, 1984) with more recent ones (Suzuki, 1991; Suzuki et al., 1993; Callegaro and Durbiano, 2003; Özkan et al., 2007).

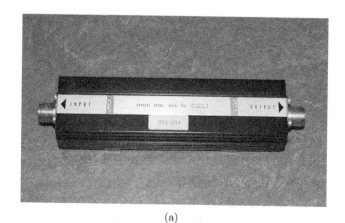

(a)

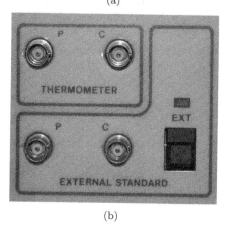

(b)

Figure 2.14
(a) An example of a 4C impedance standard, a Fluke mod. A40A 20 A current shunt. (b) Connections to two 4C impedances (above to the thermometer probe, below to an optional resistance standard) on the front panel of an ac thermometer bridge (Tinsley Consort type 5840E).

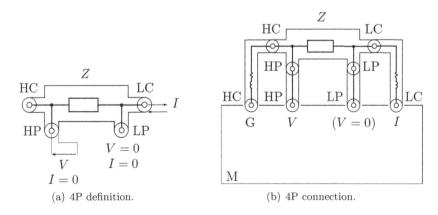

(a) 4P definition. (b) 4P connection.

Figure 2.15
(a) Four-terminal pair (4P) definition of impedance. (b) The connection of a four terminal-pair impedance standard to a meter M.

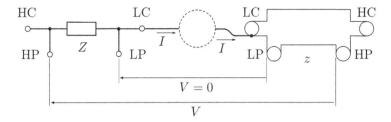

Figure 2.16
The 4P impedance of Fig. 2.15(a), under its defining condition can be redrawn by *unfolding* its screen, as when removing a surgical glove from one's hand. The unfolding shows that the internal impedance Z^* and the shield impedance z are connected in series; hence, the measured impedance is $Z = Z^* + z$.

Figure 2.17
Example of a 4P impedance, an Agilent mod. 16383A 100 pF gas-dielectric capacitor.

$V_{\mathrm{LP}} \equiv 0$, the most difficult to achieve, is often relaxed with the introduction of the so-called *combining network* (Sec. 4.6.4.3).

2.2.6   Relation with Z, Y, and S matrices

All n terminal-pair standard impedance definitions can be related to the n-rank impedance Z, admittance Y, or scattering parameter matrix Z of the standard considered a passive n-port network (Sec. 1.6.2).

Tab. 2.1 gives the relations of interest. It is important to remember that in the definition of Z and Y all port currents are considered positive if directed inward.

2.3   Change of definition

The definition of an impedance can be modified by a physical fixture or adapter, in order to be measured by a meter employing a different one. A proper adapter performs the definition conversion with a minimal change in the impedance value between the two definitions; the residual is higher with increasing frequency.

Some conversions are obvious and are performed by the connector adapters one would be forced to employ. For example, 4C to 4T conversion can be performed with coaxial-to-banana adapters of Fig. 3.20(b). In other cases, the design of an adapter or fixture require careful considerations. The general

Table 2.1

Some relations between n-terminal-pair impedance definition and the corresponding impedance Z, admittance Y, and scattering parameter S matrix elements.

Def.	Port labeling	Expression	Ref.
1P		$Z_{1P} = Z_{11}$	
2P		$Y_{2P} = -Y_{21}$	
4C		$Z_{4C} = Z_{21}$	
4P		$Z_{4P} = \dfrac{Z_{34}Z_{21} - Z_{24}Z_{31}}{Z_{31}}$	(Cutkosky, 1964)
		$Z_{4P} = 2Z_0[S_{21}S_{34} - S_{31}S_{24}] \cdot$ $\cdot [S_{31} + (S_{21}S_{32} - S_{31}S_{44} - S_{31}S_{22} + S_{41}S_{34} - $ $- S_{21}S_{32}S_{44} + S_{21}S_{34}S_{42} + S_{31}S_{22}S_{44} - $ $- S_{31}S_{42}S_{24} - S_{41}S_{34}S_{22} + S_{41}S_{24}S_{32})]^{-1}.$	(Callegaro and Durbiano, 2003)

principles for constructing an appropriate adapter have been given by Kibble (1999). Some examples are given in Fig. 2.18, 2.19.

(a) 4P resistance standard made of a Vishay 1 Ω 4T resistor embedded in a chromium-plated carved brass shell with BPO MUSA connectors.

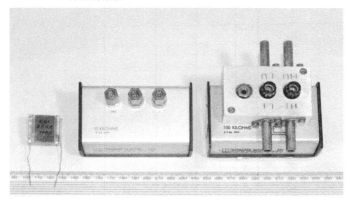

(b) ESI resistor: a 2T wirewound resistor (left) is included in a 3T standard (middle), terminated with binding posts (ESI mod. SR1), which is further transformed, with an adapter, in a 4P standard (right) terminated with BPO MUSA connectors.

Figure 2.18
Examples of impedance definition change.

Adapters and fixtures are common accessories of commercial meters. A typical fixture is the so-called *Kelvin clips*, or *Kelvin probes*, to convert the 2T definition of passive electronic components to the 4P definition of accurate impedance meters, see Fig. 2.20.

A less radical definition change is the displacement of its defining ports, to take into account finite coaxial cable lengths, connector adapters, and so on. Such displacement can be electrically modeled: the model gives calculated

Figure 2.19
Example of impedance definition change. A 4TP capacitance standard has been constructed around a 2T 10 nF electronic component (Novacap mod. 2225N103K500LZ); four coaxial cables are connected to the component and are terminated with BPO MUSA connectors on the case. Two views of the same construction are shown. See Callegaro et al. (2005) for details.

corrections that can be applied to the measured impedance. The technique is called *cable correction* or *compensation*: see Sec. 2.4 for an elementary discussion. A more complete treatment can be found in Baker-Jarvis et al. (1993).

2.4 Cable effects

As we have seen, each impedance definition has is connection diagram to the meter, which we assume is able to match the specific impedance definition conditions. One must be aware that even a perfect meter implements the definition conditions at its own connection reference plane – therefore, cables and fixtures become part of the impedance being measured. We speak of *cable corrections* if cable effects can be calculated from cable properties (either known from manufacturer's specifications, or measured on a lump of the cable type employed), and impedance readings are corrected for these effects. If, instead, the effect of the cables (or fixtures) is directly estimated on the particular setup employed, by measuring their effect on known impedance values, we speak of *cable compensations*. See Sec. 4.8. Algorithms for cable correction and compensation are often implemented in commercial instruments' firmware.

Cable corrections are possible for *n*-terminal-pair definitions and are par-

HC ⊚〉 LC
HP ⊚〉 LP

(a)

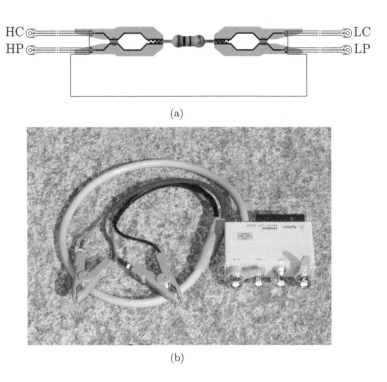

(b)

Figure 2.20
Kelvin clips. (a) The split design of the clip jaw permits to achieve 4T connection to a 2T component. The particular schematics shown achieve 4P definition. (b) In commercial Kelvin clips adapters (here an Agilent mod. 16089A, cable terminated with a compound 4-BNC connector suitable for 4P Agilent bridges), the return current conductor is often omitted.

ticularly effective for the 4P definition of impedance. For n-terminal connections, corrections have a limited accuracy.

2.4.1 Use of telegrapher's equations

The connection of a n-terminal-pair impedance to a meter include a voltage-sense and a current-sense coaxial connection (which are coincident for 1P definition). Such connections introduce propagation errors ϵ_V and ϵ_I. If the properties of the connections are known, transmission line theory (Sec. 1.6.4.4) and the resulting telegrapher's equations, Eq. (1.13) or (1.14), permit the computation of an estimate of ϵ_V and ϵ_I, in order to apply a correction to the meter reading.

For 1P, 2P, and 4C definitions, the correction is dependent on the definition and on the complex value of Z; in fact, ϵ_V and ϵ_I have a lower magnitude for high-valued impedances and 2P definition, or low-valued impedances for 4C definition.

For 4P definition, instead, the correction is happily *independent* of the impedance value. Regarding the voltage cable, we fix for convenience $x = 0$ at the meter, and $x = \ell_V$ on the 4P impedance; the meter mantains the 4P definition $I(0) = 0$ and the voltage transmission error $\epsilon_V = [V(\ell_V) - V(0)]/V(0) = +zy\ell_V^2/2$. For the current cable, taking for convenience $x = 0$ on the 4P impedance and $x = \ell_I$ on the meter, and since $V(0) = 0$[4] the current transmission error is $\epsilon_I = [V(0) - V(\ell_I)]/V(0) = -zy\ell_I^2/2$.

If cables having similar properties are employed, we have (Cutkosky, 1964; Cabiati and D'Emilio, 1975; Melcher, 1994) (Kibble and Rayner, 1984, Ch. 2.4)

$$\epsilon_{4P} \approx \frac{1}{2}\left(\ell_V^2 + \ell_I^2\right)zy. \tag{2.2}$$

For an ideal $50\,\Omega$ cable with no losses ($L = 250\,\mathrm{nHm^{-1}}$, $C = 100\,\mathrm{pFm^{-1}}$), $1\,\mathrm{m}$ cables introduce an error of 1×10^{-9} at $1\,\mathrm{kHz}$. The error increases quadratically with frequency; further increases are due to cable losses or because of small deviations from perfect four terminal-pair definition due to simplifications in the measuring circuit (Cabiati and D'Emilio, 1975).

Eq. (2.2) is embedded in some 4P commercial impedance measuring instruments for standard cable lengths, usually provided with the instruments themselves.

More complex connections, such as *four-coaxial* cables (Cabiati and D'Emilio, 1975; Melcher, 1994; Cabiati and D'Elia, 2000, 2002), permit further reduction of cable effects.

---

[4]Actually, the 4P definition condition mantained by the instrument is $V = 0$ at terminal-pair P_L, *not* at current terminal-pair C_L. However, the potential difference between the two ports is very small and, for a given current, due only to the internal construction of the standard, and independent of ℓ_I. Therefore, the analysis being carried remains valid.

3

Devices and appliances of interest in impedance measurement

CONTENTS

3.1 Generators

Impedance measuring system must include signal sources capable of energizing the impedance(s) under measurement. Below 100 MHz, the majority of modern signal generators are based on *direct digital synthesis* (DDS), see Sec. 5.2.

DDS relative frequency accuracy is the same as its clock signal generator; the internal quartz oscillator has a typical accuracy in the 10^{-6} range or better, and often can be synchronized with an external reference.

The low-power voltage output generated by DDS synthesizers can be amplified, to provide the required output level. In a laboratory environment, typical voltage and current range are below 100 V and 20 A, respectively.

In the audio frequency range, setups measuring impedances (higher than 100 Ω) are driven with voltage amplifiers, characterized by the voltage gain and a low output impedance. Setups for low impedance are often driven with *transconductance* amplifiers, which give a current output proportional to the input voltage signal with a high output impedance and up to a certain maximum voltage output capability, the *compliance*. Commercial transconductance amplifiers exist that can reach current outputs of 100 A with a bandwidth of 100 kHz.

Electromagnetic devices (transformers and inductive voltage dividers) permit to modify voltage and current output parameters with minimal power loss, and can provide electrical isolation. Generators and amplifiers employed to drive electromagnetic components must have a very low dc offset. Because of the high magnetic permeability of the core, even low dc offset currents (in the μA range) can drive the core to magnetic saturation or generate distortions and ratio errors, and such currents can be generated by voltage offsets in the the μV range in the very low dc resistance of the windings. Offset currents can be avoided with capacitive ac coupling, but resonances can occur; electronic offset control circuits (Stitt, 1990) are preferable.

Generators and amplifiers suitable for frequencies beyond the MHz range have matched outputs, i.e., the output impedance is the characteristic impedance Z_0 of the entire setup. Care must be taken if these instruments are employed to drive non-matched loads: for example, on a high-impedance load, the applied voltage is twice the displayed one.

Very high voltage or current impedance measurement setups are often driven with mains power, with the use of power transformers.

3.2   Voltage and current measurement

Impedance is defined as the ratio of voltage and current. Instruments measuring the RMS value of these quantities, with the aim of good accuracy, are called *voltmeters* and *ammeters*. Sensitive instruments measuring the same quantities down to very low ranges are called *detectors*. If the reading gives information about the voltage (current) phase respect to a given reference, the instruments are called *vector* voltmeter (ammeter) or *phase-sensitive* (or *synchronous*) detectors.

Modern electronic instruments are in essence voltage meters. If a current has to be measured, a current to voltage converter (either internal, or external, to the instrument) is employed.

The input impedance Z_{in} of real voltmeters and ammeters deviates from the ideal case ($Z_{in} = \infty$ for voltmeters, $Z_{in} = 0$ for ammeters). In many real instruments, Z_{in} can be modeled with a parallel RC equivalent; however, both R and C are frequency-dependent, and sometimes even time-fluctuating quantities (Simonson and Rydler, 1996).

In RF & MW measurements, the input impedance of the instruments is chosen to be the characteristic impedance of the measurement setup, and the distinction between voltmeters and ammeters becomes immaterial. The instruments are then called *power meters* and *detectors*.

3.2.1   RMS voltmeters

Ac voltmeters measure the root-mean-square value V_{RMS} of the periodic (with period T) voltage time signal $v(t)$, as defined in Eq. (1.7):

$$V_{RMS} = \sqrt{\overline{v^2(t)}} = \sqrt{\frac{1}{T} \int_0^T v^2(t)\,dt}. \qquad (3.1)$$

3.2.1.1   Thermal voltmeters

A thermal voltmeter implements in a direct way the definition of RMS voltage Eq. (3.1), with the Joule heating effect. The measurement is achieved by substitution in two steps: in the first, the voltage $v(t)$ of which the RMS value V has to be measured is applied to a resistor R having negligible frequency dependence. The resistor heats up because of the average dissipated power $P = \overline{v^2(t)}\,R^{-1}$ and reaches asymptotically a steady-state temperature Θ. Θ is measured with a sensitive (but not accurate) thermometer. In the second step, $v(t)$ is substituted with a dc voltage V_{dc} applied to R; the resulting steady-state temperature Θ_{dc} is measured. By definition, if $\Theta = \Theta_{dc}$, then $V_{RMS} = V_{dc}$.[1] The sequence of measurements is called *ac-dc transfer*.

The sensitivity and response time of the thermal voltmeter is dependent on the thermal properties of R and its environment (including the thermometer head), and are the result of a compromise. High thermal isolation gives higher sensitivity (since higher Θ is achieved for a given P) but increases the response time. Low thermal capacity (in practice, small physical size) increases sensitivity and speed but limit the minimum measurable signal frequency.

The accuracy of a thermal voltmeter is limited by the accuracy of the reference V_{dc} generator, the frequency dependence of its input impedance (given by R itself and the associated parasitic parameters), and the properties of the input signal conditioner (buffer amplifier and/or divider), which is necessary to unload the input signal and drive R with the appropriately scaled voltage.

Most accurate devices employed as thermal voltmeter are the single-junction (SJTC) and multi-junction (MJTC) thermal converters, Fig. 3.1. The thermometer is a thermocouple, having one (SJTC) or several (MJTC) hot junctions: the device output is a small ($\approx 1\,\mathrm{mV}$ for SJTC, tens of mV for MJTC) dc voltage. Sensitive and accurate devices are *planar* multi-junction thermal converters (PMJTC), micromachined devices where R and multijunction thermocouple are realized by thin-film litography on a thermally insulating membrane.

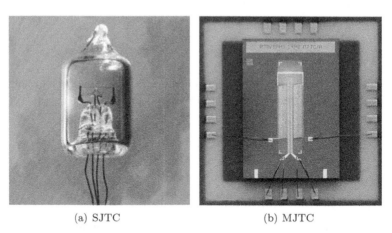

(a) SJTC (b) MJTC

Figure 3.1
Thermal converters. (a) A single-junction thermal converter in its glass vacuum-tight enclosure. (b) A planar multi-junction thermal converter, micromachined on silicon. Courtesy of Jürgen Schurr, PTB.

Somewhat less accurate, semiconductor RMS thermal converters (Fig. 3.2)

---

[1]To reduce the effect of dc voltage offsets and time drifts, more complex measurement sequences can be employed.

sense Θ with its effect on the forward voltage drop of a pn junction (Baxter, 1992). Two matched converters permit to achieve a direct continuous reading, if a feedback electronic circuit is employed (see, e.g., LT1088).

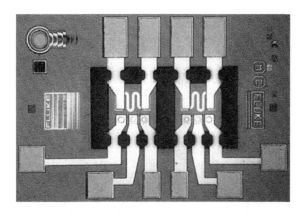

Figure 3.2
Micrography of the Fluke RMS to dc thermal converter employed in two top-class instruments, Fluke 792A ac/dc transfer standard and Fluke 5790A ac/dc voltmeter. Courtesy of Fluke Corp.

3.2.1.2 Electrostatic voltmeters

Electrostatic voltmeters implement Eq. (3.1) by sensing voltage $v(t)$ as the attractive electrostatic force F between the electrode plates of a capacitor[2], charged with $v(t)$; mechanical inertia provide the integration time constant. Common as vintage analog instruments, modern realizations have been proposed for high-voltage ac-dc transfer (Pogliano and Bosco, 2002), or as silicon micromachined RMS converters (Van Drieënhuizen and Wolffenbuttel, 1995; Bounouh and Bélières, 2011).

3.2.1.3 Analog electronic voltmeters

In analog electronic voltmeters, the definition of RMS voltage, Eq. (3.1), is implemented with an electronic circuit. The average over a period is substituted with a low-pass filtering with an appropriate time constant. Typical implementations are made with a multiplier, see Fig. 3.3, or with a log-antilog conversion (Kitchin and Counts, 1986).

---

[2]For example, in a capacitor with parallel-plane electrodes of surface area S at distance d, $F(t) = \dfrac{\epsilon S}{2d^2} v^2(t)$.

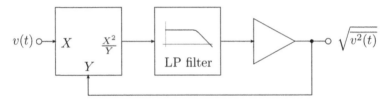

Figure 3.3
Analog RMS to dc converter employing an analog multiplier/divider in the so-called implicit computation method.

3.2.1.4 Sampling voltmeters

Sampling voltmeters measure the RMS value of $v(t)$ performing a calculation on digital samples $v[k]$ of $v(t)$ (Sec. 5.1).

If sampling (at frequency f_s) is performed synchronously (Sec. 5.3.2), a discretized version of Eq. (3.1) can be applied:

$$V_{\text{RMS}} = \sqrt{\frac{1}{n} \sum_{k=1}^{n} (v[k])^2}, \qquad n = f_s T$$

Asynchronous sampling (Sec. 5.3.3) ask for more refined calculations (see, e.g., Damle et al., 2006). For periodic waveforms, a virtual sampling frequency higher than that physically available can be achieved by *sub-sampling* (Swerlein, 1989).

3.2.2 Current to voltage converters

Modern instruments are born as voltmeters, and the current measurement function is added by implementing a current-to-voltage conversion circuitry:

resistive shunt a low-value resistor R with a low time constant, where the current I to be measured flows, see Fig. 3.4. The resulting voltage drop is measured by voltmeter V and is called *burden voltage*.

The input impedance of the voltmeter is in parallel with R. Both burden voltage and voltmeter impedance can be sources of measurement error if not properly taken into account. For this reason, shunts are not used in the measurement of very low currents, because of the corresponding high R value that would be necessary.

transresistance amplifiers an operational amplifier configured as a virtual ground current to voltage converter, see Fig. 3.5. With an ideal op amp, the transresistance gain is $G = V/I = -R$, where R is the feedback resistor value. Real transresistance amplifiers have a nonzero input impedance and

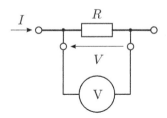

Figure 3.4
Sensing of current I with a resistive shunt R and a voltmeter V.

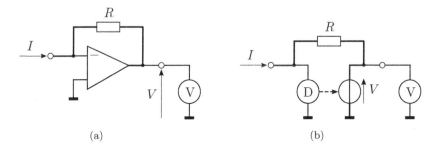

Figure 3.5
Sensing of current I with a transresistance amplifier, with feedback resistor R, and a voltmeter V. (a) Circuit diagram. (b) Functional diagram, showing the behavior of the op amp as a feedback control: detector D is nulled by acting on a voltage generator.

a frequency response dependent on the input capacitance (Pease, 2001), which can cause instability.

3.2.3   Power meters

Power meters are composed of a reading unit and interchangeable sensing heads (which can be chosen for the particular frequency band being investigated, and/or connector matching) called *power sensor*. Output is typically given in dBm.

Thermal power meters have a measurement principle similar to RMS thermal voltmeters, Sec 3.2.1.1; usually a thermistor is employed as heating sensor. *Diode* power meters rectify the input signal with a semiconductor diode.

3.2.4   Detectors

The aim of a detector is the measurement of a small signal by rejection of the background noise. Information about the frequency or phase[3] of the signal of interest must be provided.

The sensitivity achieved by a detector can be expressed in terms of its equivalent input noise spectral densities.

For low-frequency detectors, equivalent input voltage noise e_n ($V\,Hz^{-1/2}$) and equivalent input current noise i_n ($A\,Hz^{-1/2}$) are typically given; both have a roughly constant spectral density above a frequency called $1/f$ noise corner (typically in the range $10\,Hz$–$1\,kHz$). When measuring a voltage (current) source with output resistance R (conductance G), if $1/f$ noise can be neglected, the voltage and current noise floor RMS value V_n (I_n) in a given measurement bandwidth B are

$$V_n = \sqrt{[e_n^2 + (Ri_n)^2 + 4k_B R\Theta]\, B},$$
$$I_n = \sqrt{[(Ge_n)^2 + i_n^2 + 4k_B G\Theta]\, B},$$

where the last term is the Johnson noise of R (G) at temperature Θ (see Sec. 1.7.4.2).

RF & MW detectors are characterized in terms of *noise figure* (dB), the ratio between the noise power of the detector considered and that of an ideal noiseless detector (having same gain and bandwidth) measuring the same matched load; or *noise temperature*, the temperature of a matched load measured by an ideal detector which would give the same noise output power of the detector considered.

---

[3]Frequency is the first time derivative of phase; hence, phase information is more complete.

3.2.4.1 Tuned detectors

Tuned detectors are provided with a variable bandpass filter; typically, both the frequency and the quality factor (or, equivalently, the bandwidth) of the filter can be adjusted with a manual operation. The filtered signal amplitude is measured by a RMS voltmeter.

Tuned detectors have very low accuracy, since the reading is strongly dependent on the filter setting (which can drift in time). Commercially obsolete instruments, they are still presently found in high-accuracy metrology setups, since they can be battery powered and require a single coaxial connection to the measuring circuit (therefore no problem of equalization, Sec. 3.5.2) at variance with phase-sensitive detectors (Sec. 3.2.4.2).

3.2.4.2 Phase-sensitive detectors and vector voltmeters

Phase-sensitive detector, synchronous detector, homodyne detector, vector detector, and *lock-in amplifier* are basically synonimous terms. The detector is connected to the signal to be measured $v(t)$, which is amplified and filtered (with a broadband lowpass or bandpass filter) by analog electronics. The detector is also connected to the *reference* signal $v_{\mathrm{ref}}(t)$, a voltage signal (of reasonable amplitude, in the mV to V range) typically derived from the same generator energizing the measurement circuit.

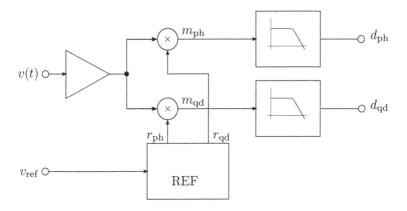

Figure 3.6
Simplified block schematics of a lock-in amplifier.

A block schematics of a phase-sensitive detector is given in Fig. 3.6. The detector derives from the reference signal v_{ref} two internal signals $r_{\mathrm{ph}}(t)$ and $r_{\mathrm{qd}}(t)$, locked in frequency and phase to v_{ref}, and in quadrature one another.

$v(t)$ is processed in two parallel channels, where it is separately multiplied by $r_{\mathrm{ph}}(t)$ and $r_{\mathrm{qd}}(t)$. Low-pass filters with cutoff frequency B extract the near-dc component of the result.[4]

The phase-sensitive detector acts as a tunable filter, with a very tight bandwidth $2B$ (which can be set down to the mHz range) which is automatically locked (hence the term *lock-in*) to the signal frequency, and which gives phase information on the measured signal. Noise rejection of commercial phase-sensitive detectors can exceed 100 dB.

Today, digital phase-sensitive detectors are commonly employed: after signal conditioning, both $v(t)$ and $v_{\mathrm{ref}}(t)$ are sampled and signal processing is carried out in the digital domain.

The phase-sensitive detector require *two* connections to the measurement system, and its power requirements typically prevent battery operation. Therefore, potential source of errors arise from interferences (induced by non-coaxiality) at the measurement frequency, which are not rejected. Equalization (Sec. 3.5.2) or isolation techniques (Callegaro et al., 1999) must be applied.

In phase-sensitive detectors, the reading accuracy is sacrificed for maximum noise rejection. *Vector voltmeters* are instruments constructed around the same principle, where accuracy specifications are optimized; in the past

---

[4]The internal representation of r_{ph} and r_{qd}, and the signal processing, can be implemented with analog or digital techniques.

Consider for simplicity only sinewave signals,

$$v(t) = V\cos(\omega t + \varphi),$$
$$r_{\mathrm{ph}}(t) = \cos(\omega_{\mathrm{ref}}t),$$
$$r_{\mathrm{qd}}(t) = \cos(\omega_{\mathrm{ref}}t + \frac{\pi}{2});$$

then the outputs of the multipliers are

$$m_{\mathrm{ph}}(t) = v(t)r_{\mathrm{ph}}(t) = \frac{1}{2}V\cos\left[(\omega - \omega_{\mathrm{r}})t + \varphi\right] + \frac{1}{2}V\cos\left[(\omega + \omega_{\mathrm{r}})t + \varphi\right],$$
$$m_{\mathrm{qd}}(t) = v(t)r_{\mathrm{qd}}(t)$$
$$= \frac{1}{2}V\cos\left[(\omega - \omega_{\mathrm{r}})t + \varphi - \frac{\pi}{2}\right] + \frac{1}{2}V\cos\left[(\omega + \omega_{\mathrm{r}})t + \varphi + \frac{\pi}{2}\right].$$

The effect of the following lowpass filters, having cutoff frequency $B \ll \omega_{\mathrm{r}}$, on signals $m_{\mathrm{ph}}(t)$ and $m_{\mathrm{qd}}(t)$ is the following:

- all terms at frequency $\omega + \omega_{\mathrm{r}}$ are filtered out;
- if $|\omega - \omega_r| \gg B$, filter outputs are null;
- if $\omega = \omega_r$, then m_{ph} and m_{qd} contain dc components, which goes unaltered through lowpass filters and can be displayed as d_{ph} and d_{qd}, proportional to the components of V phasor:

$$d_{\mathrm{ph}} = \frac{1}{2}V\cos(\varphi),$$
$$d_{\mathrm{qd}} = \frac{1}{2}V\cos\left(\varphi - \frac{\pi}{2}\right) = \frac{1}{2}V\sin(\varphi);$$

- if $|\omega - \omega_r| < B$, fluctuating d_{ph} and d_{qd} occur.

independent instruments, presently are a building block of LCR meters (Sec. 4.7).

RF & MW phase-sensitive detectors, such as those embedded in vector network analyzers (Sec. 4.11), employ *heterodyne* detection. At variance with homodyne detection, the reference frequency is shifted of a known amount, called *intermediate frequency* (IF), from the signal frequency. Signal processing after multiplying stage is carried out at IF; for example, the lowpass filters in the demodulating channels are replaced with bandpass filters tuned at IF.

3.3 Voltage and current ratio devices

3.3.1 Resistive and reactive dividers

An arrangement of two impedances permit to construct a voltage divider, Fig. 3.7(a):

$$\frac{V_1}{V} = \frac{Z_1}{Z_1 + Z_2}, \quad \frac{V_2}{V} = \frac{Z_2}{Z_1 + Z_2}, \quad \frac{V_1}{V_2} = \frac{Z_1}{Z_2},$$

or a current divider, Fig. 3.7(b):

$$\frac{I_1}{I} = \frac{Y_1}{Y_1 + Y_2}, \quad \frac{I_2}{I} = \frac{Y_2}{Y_1 + Y_2}, \quad \frac{I_1}{I_2} = \frac{Y_1}{Y_2}.$$

Resistive and reactive dividers are the basis of all-impedance bridges, Sec.

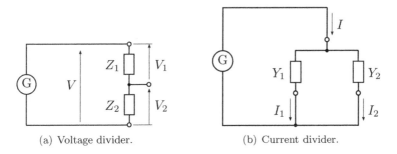

(a) Voltage divider. (b) Current divider.

Figure 3.7
Resistive or reactive dividers. (a) Voltage divider. (b) Current divider.

4.4. Their application in measurement system must take into account some effects:

- Poor input and output impedances. The device ratio is directly affected by the loading impedance Z_L. Assuming, for example, that output V_2 is loaded, a voltage divider must have $Z_2 \ll Z_L$ (in a current divider,

$Y_2 \ll Y_\mathrm{L}$); the choice fixed also the input impedance $Z_1 + Z_2$ (the input admittance $Y_1 + Y_2$), which loads the generator.

- Ratio stability is limited by the stability of the impedances composing the divider. However, if the divider is composed of similar devices (e.g., two air-dielectric capacitors, or two resistors of the same material) some rejection of environmental effects (e.g., temperature drifts) can be expected.

- Power dissipation (for resistive dividers).

- Johnson noise (for resistive dividers).

Because of these limitations, apart from special applications (for example, capacitive dividers in high-voltage systems), resistive and reactive dividers are often replaced with transformers, Sec. 3.3.2.

3.3.2 Transformers

The *transformer* is a device that transfers electrical energy from one circuit to another through inductively coupled conductors, the transformer's windings. The inductive coupling is enhanced by the transformer's *core* made of a high-permeability ferromagnetic material. A current in the *primary* winding creates a magnetic flux in the transformer's core, hence through the secondary winding, where an electromotive force is induced. Transformers employed in electrical measuring setups are called *instrument transformers* and are the subject of extensive treatises (Hague, 1936; Jenkins, 1967).

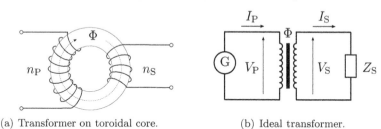

(a) Transformer on toroidal core. (b) Ideal transformer.

Figure 3.8
Transformer. (a) On the magnetic core, which carries magnetic flux Φ, primary and secondary windings having turns n_P and n_S are wound. (b) A circuit with an ideal transformer energized by generator G, which drives impedance Z_S.

Let us consider a two-winding transformer like that in Fig. 3.8. Windings P and S (primary and secondary) have turn number n_P and n_S, respectively, which concatenate the magnetic core carrying a flux Φ; let's call $t = \dfrac{n_\mathrm{P}}{n_\mathrm{S}}$ the *turns ratio*, see Sec. 1.6.4.2.

Primary winding is connected to an energizing generator G; secondary winding to an impedance Z_S (or admittance Y_S).

If all voltage drops in the windings are neglected (*ideal transformer*, see Sec. 1.6.4.2), then, in the sinusoidal regime at frequency ω, the voltage on each winding is given by Faraday law:

$$V_P = j\omega \, n_P \, \Phi,$$
$$V_S = j\omega \, n_S \, \Phi,$$

from which the transformer *voltage relation*

$$\frac{V_P}{V_S} = \frac{n_P}{n_S} = t$$

can be derived.

Φ is generated by primary and secondary currents I_P and I_S. "Ohm's law" for the magnetic core can be written $\mathcal{R}\Phi = \mathcal{M}$, where $\mathcal{R}$ is the magnetic circuit reluctance and $\mathcal{M}$ is the net magnetomotive force given by both I_P and I_S:

$$\mathcal{R}\Phi = n_P I_P - n_S I_S. \tag{3.2}$$

In an ideal transformer $\mathcal{R} \to 0$, Eq. (3.2) gives the transformer *current relation*

$$\frac{I_P}{I_S} = \frac{n_S}{n_P} = \frac{1}{t}.$$

One can easily verify that $V_P I_P = V_S I_S$ (conservation of energy).

If the secondary winding is loaded with an impedance $\dfrac{V_S}{I_S}$, see Fig. 3.8(b), then the impedance Z_P loading the generator G has the value

$$Z_P = \frac{V_P}{I_P} = t^2 \, Z_S.$$

Instrument transformers are often specified with their secondary-to-primary *voltage* ratio $k_{(V)}$ or *current* ratio $k_{(I)}$. For an ideal transformer,

$$k_{(V)} = \frac{V_S}{V_P} = \frac{1}{t}; \qquad k_{(I)} = \frac{I_S}{I_P} = t. \tag{3.3}$$

3.3.3 Real transformers

Real transformers have a finite reluctance, lossy magnetic core that require a *magnetizing current* to generate the magnetic flux Φ; windings have nonzero resistance; not the totality of magnetic flux is bounded within the magnetic core. Deviations from the ideal relationship (3.3) between voltage/current ratio and turns ratio occur.

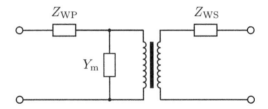

Figure 3.9
Low-frequency model of a real transformer.

A typical low-frequency equivalent circuit of a real transformer is shown in Fig. 3.9. The ideal transformer behavior is modified by a magnetizing admittance Y_m, which models the core reluctance and loss, and two series impedances Z_{WP} and Z_{WS}, which model winding resistance and *leakage inductance* (to take into account the generation of magnetic flux not bounded by the core).

According to the model, in unloaded conditions, the real voltage transformer ratio is $k_{(V)} \approx (1 - Z_{WP}Y_m)t^{-1}$.

More refined models can take into account intra- and interwinding capacitances (Binnie and Foord, 1967); the capacitive currents, not linked with the core, generate further ratio errors.

3.3.4 Transformer construction

Instrument transformers are made of enameled copper windings on a ferromagnetic core. When a winding is energized, a magnetic flux occurs. In an ideal transformer, all flux line paths flow in the core, and are therefore linked with all windings. In a real transformer some flux lines can close outside the core, linking only part of the windings, and create *leakage inductance* and corresponding transformer errors. Roughly speaking, flux lines follow the path of minimum reluctance; hence, lower transformer errors are achieved by reducing the path length, by minimizing non-ferromagnetic gaps in the core, and by increasing the permeability of the material at the operating frequency.

Audio-frequency transformer cores are made with a soft ferromagnetic metallic tape, wound in toroidal form, and encased in plastic material; core gaps are avoided, and the toroidal shape offers a minimum reluctance path to flux lines. To reduce eddy current losses (Sec. 1.10.1), the tape must have a high resistance and be very thin (typically below $100\,\mu m$); an insulation is added on the surface before winding the tape. Typical materials are permalloy-like alloys (e.g., *supermalloy*, $Ni_{79}Fe_{16}Mo_5$), or amorphous or nanocrystalline alloys. Relative permeability can be higher than 10^5.

RF transformers are wound on *ferrite* cores, powders (particles of μm size) of ferromagnetic oxides sintered in the final shape required, resulting in a material having high resistivity. Some ferrites can be employed up to GHz frequency.

Ferromagnetic behavior is limited by *saturation*, the magnetic induction value B_{sat} above which the permeability drop dramatically to values typical of paramagnetic materials. The order of magnitude of B_{sat} is 1 T, ranging from 0.2 Tdash0.5 T of ferrites to 2 T of iron alloys. Saturation limits the maximum magnetic flux a given core can withstand, and (given the number of turns and the working frequency) the maximum working voltage of a winding.

The stages of construction of a multitap transformer are shown in Fig. 3.10.

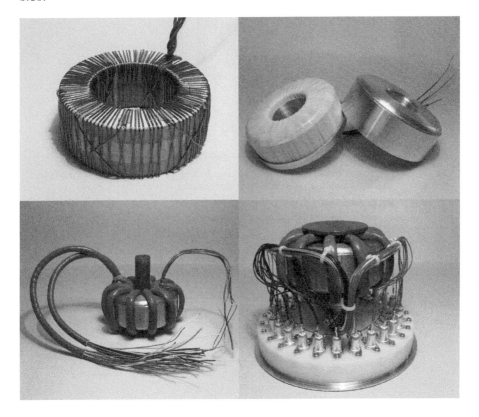

Figure 3.10
The construction stages of a transformer having a primary winding and a secondary winding with 20 taps. (upper left) The primary winding on a toroidal core. (upper right) An electrostatic shield made of copper, before being closed over the primary winding. (lower left) Rope secondary winding. (lower right) Connection of the secondary winding to tap connectors, arranged in circular fashion.

3.3.5   Two-stage transformers

The ratio error of the voltage transformer of Fig. 3.9 is caused by the voltage drop $V_P' = V_P - E_P$ (due to stray parameters Z_{WP} and Y_m) between primary voltage V_P and the electromotive force E_P created by core flux Φ.

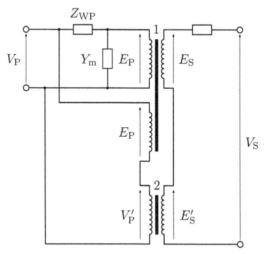

(a) Arrangement of two voltage transformers having equal turns ratio. Voltage ratio error of transformer 1 is corrected by transformer 2.

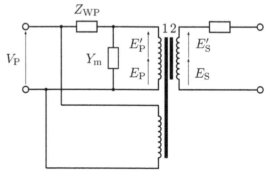

(b) Two-stage voltage transformer. The arrangement of (a) is condensed in a single device having two cores, 1 and 2.

Figure 3.11
(a) Description of the two-stage principle with two separate transformers. (b) Two-stage voltage transformer.

Such ratio error can be corrected to a certain extent with the circuit in Fig. 3.11(a). On core 1 the original transformer has, in addition to the primary winding (with n_P turns) and the secondary winding (with n_S turns), an additional sensing winding with n_P turns. Such winding can sense electromotive force E_P. The difference V_P' can feed a second transformer, wound on a second core 2, having the same n_P/n_S turn ratio of that on core 1; the resulting secondary voltage E'_S is the correction voltage, injected in series to E_S. Although also transformer 2 has a ratio error, this is less significant because $V'_P \ll V_P$.

The *two-stage* or *double-stage voltage transformer* (Deacon and Hill, 1968) design collapses the two transformers of 3.11(a) in a single unit having two cores and three windings, as shown in Fig. 3.11(b). Core 1 is wound with the *magnetizing* winding; then cores 1 and 2, stacked together, are wound with primary and secondary windings, as schematically shown in Fig. 3.12(a).

Both primary and magnetizing winding are excited with primary voltage V_P. The magnetizing winding generates the electromotive force E_P in the primary winding, which must now sustain in core 2 only the smaller electromotive force E'_P, linked to voltage difference V'_P. On the secondary side, the electromotive forces E_S (due to core 1) and E'_S (due to core 2) are in series.

A dual arrangement is employed in the *double-stage current transformer* as shown in Fig. 3.12(b).

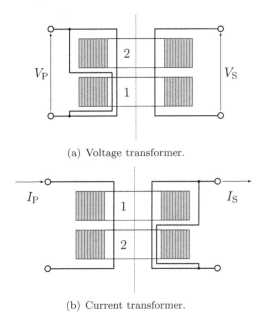

(a) Voltage transformer.

(b) Current transformer.

Figure 3.12
Two-stage transformers: simplified cross-section, 1-turn windings on two toroidal cores.

3.3.6 Inductive dividers

Transformers having more than one secondary winding, or having *taps* on winding turns, can be constructed. *Autotransformers*, or *inductive dividers* have the primary and secondary windings having a fraction of turns in common, by using *taps* on the conductor which composes the turns. An example of inductive voltage divider (IVD) is shown in Fig. 3.13.

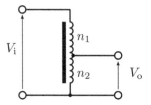

Figure 3.13
Inductive voltage divider. The single winding is divided into two sections having n_1 and n_2 turns by a tap, where the output voltage is taken.

The (output-to-input) voltage ratio of the ideal IVD is

$$k = \frac{V_\mathrm{o}}{V_\mathrm{i}} = \frac{n_2}{n_1 + n_2}.$$

As happens in real transformers, the voltage ratio of real IVDs is different from that calculated from turns numbers. However, it is found that IVD error is smaller than the error of a transformer of similar construction and same nominal ratio; a good IVD can have a voltage ratio error below 1×10^{-7} at audio frequency. In addition, such error is extremely stable with time or environmental conditions (Briggs et al., 1993). Inductive dividers are therefore preferred to transformers as ratio elements in impedance meters.

Similarly to transformers, IVDs can operate also as current dividers.

The calibration of IVDs can be performed by comparison with a reference. The self-calibration can be performed by step-up procedures, the so-called *bootstrap* or *straddling* methods (Hill and Deacon, 1968a,b; Hanke, 1989; Callegaro et al., 2003); or by embedding the IVD in a transformer ratio bridge (Sec. 4.5), and employing the *permuting capacitor* technique (Cutkosky and Shields, 1960; Awan et al., 2000).

3.3.7   Decade dividers

The ratio of an IVD can be varied by connecting the output to different available taps. To achieve greater resolution in the ratio selection, two or more IVDs can be connected in a cascade. A typical arrangement is the *Kelvin-Varley* divider, shown in Fig. 3.14 (Hill and Miller, 1962).

In practical constructions of variable IVDs, usually more than one decade is wound on the same core, see Fig. 3.15(a) and 3.16. To permit ratio selection during operation, a particular arrangement of the decade switches wiring permits to avoid the temporary disconnection of windings and the consequent discharge of its large inductance, which create voltage spikes.

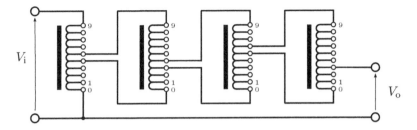

Figure 3.14
A variable inductive voltage divider, with four decades in Kelvin-Varley connection.

3.3.8 Sum-difference transformers

The special voltage transformer of Fig. 3.17 has two cores, 1 and 2, each wound with a primary winding having n_1 and n_2 turns, respectively. A secondary winding with n_Δ turns is wound on both cores. If the primaries are fed with voltages V_1 and V_2, the arrangement permits to obtain the linear combinations $\Delta V = \pm \dfrac{n_\Delta}{n_1} V_1 \pm \dfrac{n_\Delta}{N_2} V_2$. If $n_\Delta = n_1 = n_2$, the transformer is known as a sum-difference voltage transformer: $\Delta V = \pm V_1 \pm V_2$.

The working principle of the sum-difference transformer is the same as a circuit composed of two independent transformers, where the secondaries are put in series. However, the arrangement of Fig. 3.17 is more compact and permits to reduce secondary winding stray parameters. To improve ratio accuracy, a two-stage realization (Sec. 3.3.5) employing four cores can be considered (Cabiati and Pogliano, 1987; Pogliano et al., 2011).

3.3.9 Feedthrough transformers

In feedthrough transformers, Fig. 3.18 one of the windings is given by a conductor crossing the magnetic circuit, thus realizing a single turn. The other winding, having N turns, is normally wound around the core. The typical use of feedthrough transformers in impedance measurement circuitry is twofold:

injection transformer, as an isolated source of a small electromotive force. The primary winding is fed with voltage E, an electromotive force $N^{-1} E$ is injected in the mesh, which includes the conductor realizing the secondary winding. For improved accuracy, two-stage realizations are employed.

detection transformer, current I flowing through the single-turn conductor magnetizes the core and induces an output voltage $\omega M I$ (where M is the windings' mutual inductance) at the primary winding; V can thus be sensed by a detector. Typically, the detection transformer is employed as a

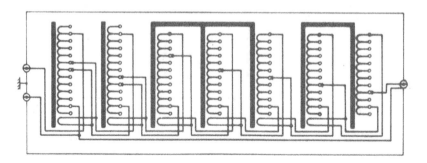

(a) The seven decades are wound on four separate cores: decade 10^{-1} on core 1, decade 10^{-2} on core 2, decades 10^{-3}, 10^{-4}, 10^{-5} on core 3, decades 10^{-6} and 10^{-7} on core 4.

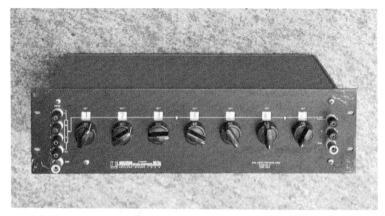

(b) Front panel, showing the seven rotary switches and connections.

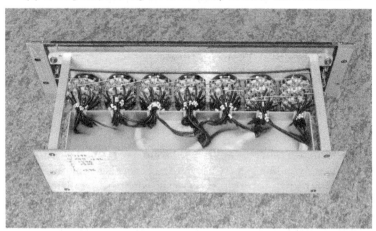

(c) Inner view, showing the connections to rotary switches. The windings are casted in resin.

Figure 3.15
Seven-decade inductive voltage divider ESI mod. DT72A.

Figure 3.16
The inner construction of a six-decade IVD employing three separate cores.
Decades 10^{-1} (300 turns) and 10^{-2} (30 turns) are wound on core 1, decade
10^{-3} (200turns) and 10^{-4} (20 turns) on core 2, decades 10^{-5} (100 turns) and
10^{-6} (10 turns) on core 3.

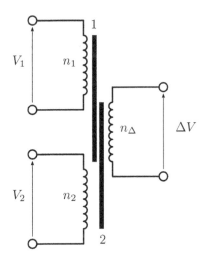

Figure 3.17
Sum-difference voltage transformer.

zero detector, hence precise knowledge of the transformer transimpedance is not important.

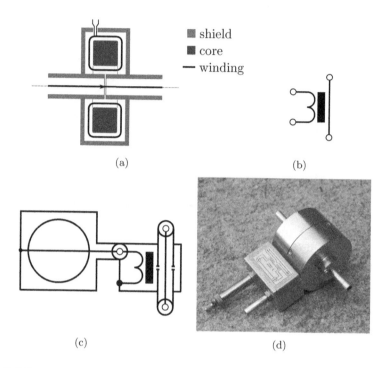

(a)

■ shield
■ core
— winding

(b)

(c)   (d)

Figure 3.18
Feedthrough transformer. (a) A section. (b) Non-coaxial symbol. (c) Coaxial injection transformer connected to a voltage source. (d) External view of a feedthrough transformer, terminated with BPO MUSA connectors.

3.4   Cables and connectors

Electrical wiring for laboratory impedance measurements is usually not permanent, and performed with cables terminated with plug connectors on instruments provided with panel socket connectors.

3.4.1   Cables

Power-frequency impedance measurements can be performed with single wires, which isolation and cross section must sustain the voltage and current to be carried. To reduce the parasitic inductance of the measurement circuit (and consequently both its sensitivity to and generation of electromagnetic interference), wires can be coupled in *twisted pairs*, possibly with an electrostatic shielding.

Impedance measurements at audio or radio frequency are performed with flexible coaxial cables; microwave measurement can benefit the use of semi-rigid cables. Geometrical dimensions and material properties influence the electrical properties of the cable as a transmission line, see Sec. 1.6.4.4. For a lossless cable (perfect conductor and dielectric) of cylindrical geometry, the shunt capacitance C and series inductance L per unit length, and the characteristic impedance Z_0, are given by

$$C = \frac{2\pi\epsilon_0\epsilon_r}{\log\left(\frac{D}{d}\right)}, \qquad L = \frac{\mu_0\mu_r}{2\pi}\log\left(\frac{D}{d}\right), \qquad Z_0 = \frac{1}{2\pi}\sqrt{\frac{\mu_0\mu_r}{\epsilon_0\epsilon_r}}\log\left(\frac{D}{d}\right);$$

where D is the inner diameter of the outside conductor, d is the outer diameter of the inner conductor, ϵ_r and μ_r the relative permittivity and permeability of the insulator ($\mu_r \approx 1$).

The materials employed influence also the cable series resistance R and the shunt conductance G per unit length, and therefore the cable *loss*, which increases with frequency because of skin effect in conductors and hysteresis loss in dielectrics.

Commercial coaxial cables have $Z_0 = 50, 52, 75$, or $93\,\Omega$, with a majority of cables having $Z_0 = 50\,\Omega$. For high-frequency applications, the characteristic impedance of the cable must correspond to that of instruments and standards employed. For low-frequency impedance, the measurement system can benefit of either a low L or a low C cable choice, in order to minimize cable corrections (Sec. 2.4).

Sensitive high-value impedance measurements benefit of the use of *low-noise* cables, where a semi-conductive lubrication layer between conductors and insulation reduce the buildup of triboelectric charges generated by mechanical vibrations.

Triaxial cable, Fig. 3.19 can be employed in specialized impedance measurements, when the outer connector of the coaxial pair must be kept at a potential considerably different from the environment, such as in guarding methods (Sec. 3.5.1).

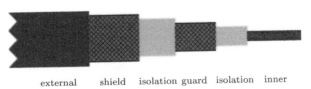

external   shield   isolation guard   isolation   inner

Figure 3.19
Schematic view of a triaxial cable.

3.4.2   Connectors

3.4.2.1   Single-wire connectors

Several impedance standards and instruments are provided with *binding post* connectors. Although binding posts can accept also bare wires, typical connections are performed with wires terminated with terminated with *banana plugs* or *spade terminals*, see Fig. 3.20.

(a) Standard (4 mm diameter) banana plugs on isolated conductor.

(b) BNC to double banana plug adapter. The banana corresponding to the BNC outer conductor is marked.

(c) Spade terminal on a high-current (10 A) cable.

(d) Binding posts on the top panel of an impedance standard. The binding post can accept banana plugs (single or double), spade terminals, even bare wire.

Figure 3.20
Some single-wire connections.

3.4.2.2   Coaxial connectors

Several types of coaxial connectors are employed in impedance measurement. A partial list is given in Tab. 3.1. Low-frequency, primary impedance metrology appreciates the MUSA connector, because of the ease of mating action and the very low and repeatable contact resistance (Callegaro et al., 2002)

Table 3.1
Some connectors employed for impedance measurements. BNC, N, SMA images
courtesy of Huber+Suhner AG, Switzerland.

name	lock	photo	max f	notes
BNC	male-female, twist-lock		4 GHz	*Bayonet Neill-Concelman.* 50 Ω and 75 Ω available. Poor mechanical stability.
GR874	hermaphrodite, threaded		8.5 GHz	*General Radio mod. 874.* Obsolete, but still found on several primary impedance standards.
N	male-female, threaded		18 GHz	*Neill.* 50 Ω and 75 Ω available
APC-7	hermaphrodite, threaded		18 GHz	*Amphenol Precision Connector.* Also called *7 mm.* 50 Ω.
SMA	male-female, threaded		26 GHz	*Subminiature A.* 50 Ω. SMA-like connectors, like the 3.5 mm can go up to 34 GHz.
MUSA	male-female, slide-in			*Multi-User Steerable Array,* also BPO (*British Post Office*). 50 Ω historical, 75 Ω today. Very low and stable contact resistance. Originally developed for patch bays.

3.5   Shielding

Electrical shielding is the technique of enclosing all components (devices and connecting conductors) of a measurement circuit inside a closed metal surface that is at low potential with respect to the environment. In this way, electric fields (and the corresponding displacement currents) from the environment to the circuit, or from one part of the circuit itself to another part, are intercepted by the shield.

Magnetic shielding is the technique of enclosing sensitive components within a three-dimensional enclosure made of a ferromagnetic material having a high permeability (soft magnetic material). A typical material employed is *mu-metal*, having a relative permeability that can exceed 100 000. Environmental magnetic fields follow a low-reluctance path through the enclosure, and their intensity within the enclosure is strongly reduced. Magnetic shielding can easily become ineffective: a high-field transient can permanently magnetize the material (thus strongly reducing its permeability); mechanical shocks or plastic deformations can irreversibly modify the material properties. For this reason, magnetic shields are machined in the final shape and then processed by high-temperature annealing.

For variable fields, a certain degree of magnetic shielding is given also by conducting enclosures, because of shielding effect given by parasitic currents (Lenz's law). The effect can be quantified by the penetration depth, Sec. 1.10.1. For example, in aluminum ($\mu \approx \mu_0 = 1.26 \times 10^{-6}\,\mathrm{H\,m^{-1}}$, $\sigma = 37.8\,\mathrm{MS\,m^{-1}}$) the penetration depth at 1 kHz is 2.5 mm.

A very peculiar magnetic shielding is given by *superconductors*: if the external magnetic field is lower than a certain value characteristic of the material called the *critical field*,[5] superconductors manifest the *Meissner effect*, the complete exclusion of magnetic induction (that is, perfect diamagnetism) with the production of surface-shielding currents. A volume enclosed in a superconducting surface is therefore completely shielded by external dc or ac magnetic fields; if the volume is multiply connected, e.g., in a torus, a magnetic flux can be perfectly confined within.

3.5.1   Guarding

Guarding is a particular shielding method, see Fig. 3.21. The guard shield is driven by a potential similar to the conductor being guarded (or, in multi-conductor arrangements, to one of the conductors being guarded). The conductor-to-environment admittance is splitted in a conductor-to-guard and guard-to-environment

---

[5]For Type I superconductors, like Pb. For Type II superconductors, it's called *lower critical field*.

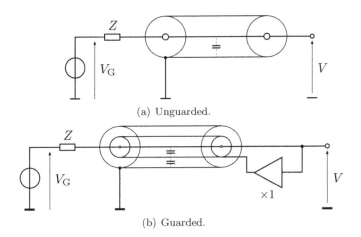

(a) Unguarded.

(b) Guarded.

Figure 3.21
Guarding principle. (a) A voltage generator with output impedance Z drives
a coaxial cable with voltage V_G; the inner-to-outer cable capacitance loads
the generator, and output voltage $V \neq V_G$. (b) The same voltage generator
drives a triaxial cable; the guard is driven with a low-impedance generator to
output voltage V (here with a unity-gain buffer amplifier), which can sustain
the guard-to-low capacitive currents. Since inner and guard conductors are at
the same potential, no current flows through the inner-to-guard capacitance,
and $V \sim V_G$.

3.5.2 Coaxiality

The principle of *coaxiality* (Espenschied and Affel, 1931) is shown in Fig. 3.22. The coaxial transmission structure (coaxial cable) is composed of an inner and an outer conductor, having the same axis. The outer conductor acts as shield, and intercepts electrical fields (and corresponding displacement currents) from the inner conductor, or from the environment. The structure is connected to electrical devices (generator G in Fig. 3.22) such that opposite currents of equal magnitude I flow in the inner and the outer conductors. In this way, the magnetic field associated with I is completely enclosed in the space within the two conductors, and none develops in the environment. Because of the reciprocity theorem, the same transmission structure is immune to external sources of interference.

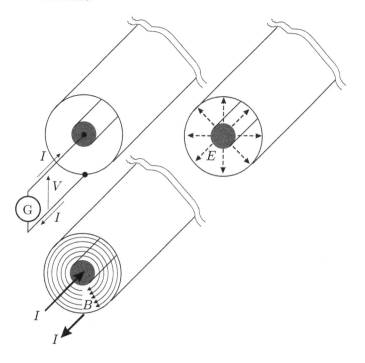

Figure 3.22
Coaxial structure, showing that both electric and magnetic field are confined in the dielectric between the inner and outer conductor.

Although perfect coaxiality is impossible to achieve in practice, the use of common flexible coaxial cables, the enclosure of all electrical devices in a closed conducting enclosure, and the use of coaxial connectors constitute an excellent approximation. Electrical circuits having a tree structure, if connected to ground at a single point, are automatically coaxial; when instead the circuit contains meshes, multiple current paths require *equalization*.

For example, the circuit of Fig. 3.23(a) is shielded but not necessarily coaxial.

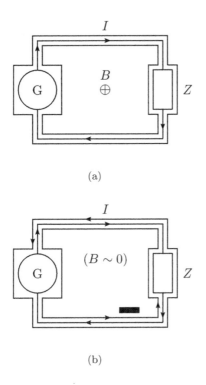

(a)

(b)

Figure 3.23
A shielded circuit with one mesh, including a generator G and impedance Z. (a) Nonequalized condition: inner current I flows through the inner conductor mesh, but current through outer conductor mesh is not under control, and at low frequency can be assumed to be negligible. The circuit is noncoaxial and magnetic field B appear in the circuit environment. (b) A current equalizer creates a current I in the outer conductor mesh flowing in the reverse direction; magnetic field in the circuit environment is suppressed.

At radio or microwave frequency, equalization is automatically achieved; inner and outer conductors are magnetically coupled, and the mutual reactance generate a voltage in the outer mesh which (by Lenz's law) drive the equalization current I. The magnetic field in the space where the mesh is located is nulled, minimizing the electromagnetic energy. At power or radio frequency, the inner-outer mutual reactance (proportional to frequency) must be raised by simple devices called *current equalizers* or *chokes*, composed of a

Figure 3.24
Coaxial equalizer, or *choke*. The coaxial cable is wound around a high-permeability toroid (here a nanocrystalline tape toroid), and sandwiched between two plastic flanges for mechanical stability. The toroid carries an additional single-wire winding, terminated with a MUSA connector on the top flange. This winding can be employed in tests of the efficiency of the equalizer: it can be shorted (this suppresses the equalizer without need to disconnect it) or connected to a voltage detector to sense the flux in the toroid.

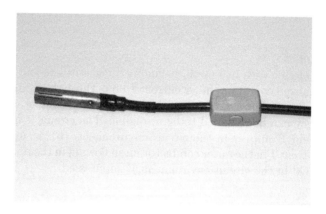

Figure 3.25
High-frequency coaxial equalizer. A ferrite bead, in the form of a hollow tube, is clamped on a coaxial cable, giving a one-turn coaxial equalizer, suitable for high-frequency signals.

Figure 3.26
Two typical symbols for current equalizers found in the literature.

few turns of coaxial cables wound on a high-permeability ferromagnetic core, see Figs. 3.24 and 3.25. For better performance, *active* coaxial equalizers can be employed (Homan, 1968). One (and only one!) current equalizer for each closed mesh in the measurement circuit has to be employed. In circuit drawings, the equalizer symbol is often found as in Fig. 3.26.

The circuit of Fig. 3.23(a) has one mesh, so it can made coaxial with one current equalizer, as in Fig. 3.23(b).

A recent book by Awan et al. (2011) has been devoted to the issue of coaxiality in measurement circuits.

4

Common practice methods

CONTENTS

4.1   *I-V* method

The simplest approach to the measurement of impedance is a direct application of its definition (Sec 1.4), resulting in an implementation of the *I-V method.*

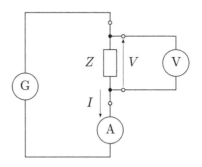

Figure 4.1
I-V measurement of a four-terminal (4T) impedance Z with a voltmeter V and an amperemeter A.

As shown in Fig. 4.1, the impedance Z is energized by a generator G; the current through Z is measured by the ammeter A; the corresponding voltage drop V on Z is measured with a voltmeter V. The impedance magnitude is indirectly measured as $|Z| = \dfrac{V}{I}$. If V and A are phase-sensitive instruments (a vector voltmeter and a vector ammeter, see Sec. 3.2.4.2) and give also phase readings φ_V and φ_I relative to an arbitrary phase reference, the complex impedance

$$Z = \frac{V}{I} \exp\left[j\left(\varphi_V - \varphi_I\right)\right] \tag{4.1}$$

can be measured.

The method can be implemented to achieve the particular impedance definition (Ch. 2) of interest. For example, Fig. 4.2 shows the application of the *I-V* method to a two-port (2P) impedance.

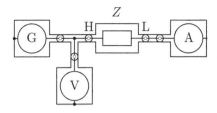

Figure 4.2
I-V measurement of a 2P impedance.

4.2 Two-voltage methods

Two-voltage methods implement the ammeter function needed in the *I-V* method with a current-to-voltage converter (Sec. 3.2.2) and an additional voltage reading.

The main advantage of two-voltage methods derives from the measurement model of the *I-V* method, Eq. (4.1), where amplitude ratios and phase differences occur. If a single vector-reading instrument for measurement of both V and I (by properly switching it between two positions), its absolute amplitude and phase measurement accuracy is not important, and only its linearity and short-term stability enter the uncertainty budget.

4.2.1 Vector ratiometer

The current-to-voltage converter of the *I-V* method can be implemented with a reference impedance (Sec. 3.2.2), chosen such that $|Z_1| \sim |Z_2|$. The resulting circuit can be called a vector ratiometer, Fig. 4.3; its model equation is

$$\frac{V_1}{V_2} = \frac{Z_1}{Z_2}.$$

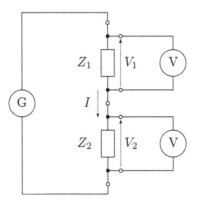

Figure 4.3
Vector ratiometer, for the comparison of two 4T impedances Z_1 and Z_2.

The measurement can be performed with a single-vector voltmeter, switched between Z_1 and Z_2. Care must be taken to avoid errors caused by the loading of V on V_1 and V_2, in particular when a single switched voltmeter is employed.

Overney and Jeanneret (2011) give a recent implementation of a vector ratiometer for 4P impedance comparisons, employing a synchronous-sampling vector voltmeter (Sec. 5.3.2).

4.2.2   Self-balancing bridge

If the current-to-voltage converter is a transresistance amplifier (Sec. 3.2.2), the resulting circuit, Fig. 4.4, can be considered a self-balancing *bridge* (Sec. 4.4).

The self-balancing bridge method is the basis of LCR impedance meters (Sec. 4.7).

4.3   Three-voltage method

The rationale of the *three-voltage method* is to overcome the relatively low accuracy of vector voltmeters, and rely on the superior performance of ther-

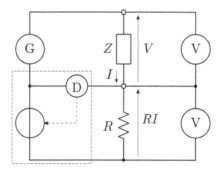

Figure 4.4
An implementation of *I-V* method for the measurement of the impedance Z by employing an ammeter with a transresistance input stage, here redrawn as a self-balancing bridge. The op amp of the transresistance amplifier is displayed by its functional diagram (Sec. 3.2.2), as splitted in a sensing detector D and a controlled voltage generator. The feedback resistor R acts as the impedance standard, to which Z is compared.

mal RMS voltmeters (Sec. 3.2.1.1), which, however, do not give any phase information.

As the name of the method name suggests, three RMS voltage measurements are involved. The idea behind the method is that the graphical representation of the phasors of each voltage can be arranged in a closed triangle. The RMS value of each voltage corresponds to the geometrical length of the corresponding triangle side. Geometry theorems link the properties of the triangle: for example, given the side lengths (phasor magnitudes) the angles at vertices (phase differences between phasors) are completely determined.

4.3.1 Principle

The principle of impedance measurement with the three-voltage method[1] is illustrated in Fig. 4.5. The impedance standard $Z_S = R_S + jX_S$ and unknown $Z_X = R_X + jX_X$ 2T impedances are connected in series. When current I flows in the mesh, voltage drops $V_S = Z_S I$ and $V_X = Z_X I$ develop on Z_S and Z_X; a total voltage $V = V_S + V_X$ develops on the series. The three RMS voltage values $|V_X|$, $|V_S|$, and $|V|$ are measured.

By applying elementary geometry to the phasor triangle $\{V_S, V_X, V\}$, it

---

[1]The three-voltage method has been employed for long time in power measurements (Walker, 1941; Marzetta, 1972; Waltrip and Oldham, 1997); its application to impedance measurement is due to Cabiati (Cabiati and Bosco, 1992; Cabiati et al., 1994; Muciek, 1997; Muciek and Cabiati, 2006).

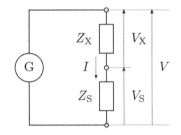

(a) The basic electrical circuit.

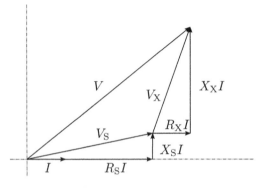

(b) Phasor diagram.

Figure 4.5
Principle of the three-voltage method.

can be shown (Muciek and Cabiati, 2006) that

$$R_X = \alpha_B R_S - \alpha_A X_S,$$
$$X_X = \alpha_A R_S + \alpha_B X_S, \tag{4.2}$$

where α_A and α_B are functions of ratios of $|V_X|$, $|V_S|$, and $|V|$:

$$\alpha = \frac{|V|}{|V_S|},$$

$$\alpha_X = \frac{|V_X|}{|V_S|},$$

$$\alpha_A = \frac{1}{2}\sqrt{\left[1 - (\alpha - \alpha_X)^2\right] \cdot \left[(\alpha + \alpha_X)^2 - 1\right]},$$

$$\alpha_B = \frac{1}{2}\left(\alpha^2 - \alpha_X^2 - 1\right). \tag{4.3}$$

The three-voltmeter method ask for some care because of an intrinsic ambiguity: given the three triangle sides, *two* triangles can be constructed, as shown in Fig. 4.6. Eq. (4.3) is valid when $\arg(Z_X) > \arg(Z_S)$, like in the diagram of Fig. 4.5(b).

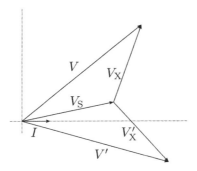

Figure 4.6
The intrinsic ambiguity embedded in the three-voltage method: two triangles having same sides $|V_S|$, $|V_X| = |V_X'|$ and $|V| = |V'|$ can be constructed.

4.3.2 Application of the three-voltage method

The three-voltage method is of interest when Z_S and Z_X are of different kinds, e.g., when calibrating an inductor in terms of a resistor. In such cases, the ambiguity of Fig. 4.6 is easily solved.

When measuring impedance asking for a definition more complex than the 2T one, the application of the three-voltage method require the solution of two issues:

measurement of V Accurate voltmeters have a limited common-mode rejection ratio. If the impedances under comparison are configured as 3T, 2P, or 4P standards, the measurement of V_S and V_X pose no problem, but with such definitions V has a high common-mode voltage respect to screen potential.

connection voltage drops 4T and 4P definitions must deal with voltage drops in the current circuit.

Modifications of the original principle help to solve such difficulties.

Inductive voltage divider Equations (4.2) and (4.3) are based on a triangle constructed on the phasor linear combination $V = V_S + V_X$. Other linear combinations can be chosen: for example, the voltage difference $V_m = (V_X - V_S)$, which gives (Fig. 4.7) the triangle $\{V_S, V_m, V_X\}$.

As shown in Fig. 4.8(a), the voltage $V_M = \frac{1}{2}V_m$ can be obtained with the help of an inductive voltage divider, having a tap in the middle of

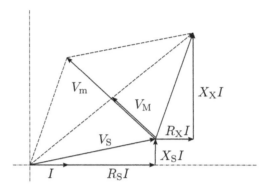

Figure 4.7
Same phasor diagram of Fig. 4.5(b), where difference voltages $V_m = V_X - V_S$
and $V_M = \frac{1}{2}(V_X - V_S)$ are displayed.

the winding (ratio 0.5); its magnetizing current is provided by the circuit generator. The measurement of V_m is not affected by common-mode difficulties.

Combined with the use of *compensations* (Sec. 4.6.3.3), such solution has been employed in an automated measurement system for 4P impedance comparisons with a base accuracy better than 1×10^{-5} (Callegaro and D'Elia, 2001; Callegaro, 2007; Callegaro et al., 2009).

Sum-difference transformer, Sec. 3.3.8, permits to obtain in a direct way either V or V_m referenced to any potential. The method can be applied directly to 4T low-valued impedance, see Fig. 4.8(b), where the transformer loading can be neglected (to reduce loading, a two-stage sum-difference transformer can be considered).

If the voltage difference $V_m = V_X - V_S$ or the half-difference $V_M = V_m/2$ is measured, Eq. (4.3) is modified to

$$\alpha_m = \frac{|V_m|}{|V_S|} = 2\frac{|V_M|}{|V_S|},$$

$$\alpha_X = \frac{|V_X|}{|V_S|},$$

$$\alpha_A = \frac{1}{2}\sqrt{\left[1 - (\alpha_m - \alpha_X)^2\right] \cdot \left[(\alpha_m + \alpha_X)^2 - 1\right]},$$

$$\alpha_B = \frac{1}{2}\left(1 + \alpha_X^2 - \alpha_m^2\right). \tag{4.4}$$

Since the calculations of Eq. (4.2), (4.3), and (4.4) involve only RMS volt-

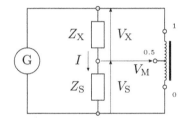

(a) IVD solution to measure V_M.

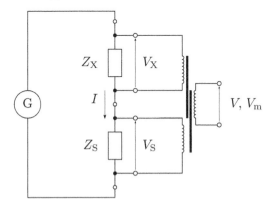

(b) Difference transformer solution, to measure either V or V_m.

Figure 4.8
(a) an IVD or (b) a difference transformer can help to solve the problem of measurement of $|V|$ in the basic scheme of Fig. 4.5(a).

age *ratios*, when a single voltmeter is employed for all three voltage measurements (by appropriate manual or automated switching), only its linearity and short-term stability, and not its absolute accuracy, has to be considered in the uncertainty budget. The expression of the uncertainty of the method is particularly involved (Muciek and Cabiati, 2006), because of the nonlinear nature of Eq. (4.3) and (4.4); the application of Monte Carlo method (GUM Suppl. 1), however, is straightforward (Callegaro and D'Elia, 2001).

4.4   Bridges

The *bridge* is a measuring method invented by Christie (1833) and popularized by Wheatstone (1843). The basic arrangement, called the *Wheatstone bridge*, is shown in Fig. 4.9. Impedances under comparison Z_1, Z_2, Z_3, Z_4 are called *branches* or *legs* of the bridge. Generator G drives the bridge; D is a voltage or current detector.

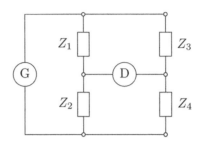

Figure 4.9
The Wheatstone bridge.

The bridge is in *equilibrium* if D reads zero: then, by Ohm's law,

$$\frac{Z_1}{Z_2} = \frac{Z_3}{Z_4}. \tag{4.5}$$

The bridge measures one of the impedances (e.g., Z_4), the *unknown*, in terms of the others (Z_1, Z_2, and Z_3), the *standards*. The equilibrium is achieved by having one (or more) of the three standards in the form of variable components (e.g., decade boxes), and by modifying the complex value of such standard, until D reads zero.

A classification (Ferguson, 1933) divides the bridges in *ratio bridges* if the adjustable standard has a node in common with the unknown (Z_2 or Z_3 in Fig. 4.9) and in *product bridges* if the adjustable standard is on the opposite side (Z_1) with respect to the unknown.

If the detector and source positions are switched, the bridge is still balanced; this *reciprocity* property is common to all linear bridge networks.

Wheatstone-like bridges have different names, depending on which kind of impedances constitute the branches. Some are collected in Tab. 4.1; a more extensive taxonomy is given by Hague (1971, Ch. IV).

The all-impedance bridge of Fig. 4.9 can be reconsidered as:

- a *voltage ratio bridge*, Fig. 4.10(a). Impedances Z_1 and Z_2 constitute a voltage divider (Sec. 3.3.1), which outputs voltages E_1 and E_2 having a known ratio. The bridge equation is

$$\frac{E_1}{E_2} = \frac{Z_3}{Z_4};$$

- a *current ratio bridge*, Fig. 4.10(b). Impedances Z_1 and Z_3 constitute a

Name	Scheme	Relation(s)	Calibration of
Wheatstone		$R_4 = \dfrac{R_2\,R_3}{R_1}$	pure resistance
De Sauty		$C_4 = C_3\dfrac{R_1}{R_2}$	pure capacitance
Wien		$C_4 = C_3\dfrac{R_1}{R_2}$ $\qquad R_4 = \dfrac{R_2\,R_3}{R_1}$	lossy capacitance
Maxwell		$L_4 = R_2 R_3 C_1$ $\qquad Q_4 = \omega R_1 C_1$	lossy inductance

Table 4.1
All-impedance bridges.

current divider, which outputs currents I_1 and I_3 having a known ratio. The bridge equation is

$$\frac{I_1}{I_3} = \frac{Z_4}{Z_2}.$$

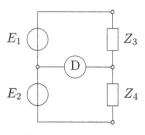
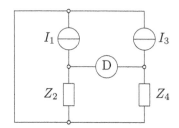
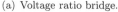

(a) Voltage ratio bridge.   (b) Current ratio bridge.

Figure 4.10
(a) Voltage and (b) current ratio bridges.

The interest in redrawing Fig. 4.9 as in Fig. 4.10 is that *any* voltage or current ratio device (Sec. 3.3) can be employed to construct a viable bridge. Very effective devices in the audio frequency range are inductive dividers.

4.4.1   Directional bridge

A particular version of Wheatstone bridge, called *directional bridge*, is employed as a directional coupler in network analyzers, Sec. 4.11 (Yonekura and Jansons, 1994). Let Z_0 be the characteristic impedance of the analyzer. The directional bridge has $Z_1 = Z_2 = Z_3 = Z_0$ (see Fig. 4.9); the instrument test port is where impedance Z_4 is connected. The directional bridge works in unbalanced condition, and the mismatch at the test port is computed from the detector reading V_D. For example, the equilibrium condition $V_D = 0$ (that is, $Z_4 = Z_0$) corresponds to zero reflection ($S_{11} = 0$) at the test port.

4.4.2   T-bridges

Bridges of Tab. 4.1 require isolation between source and detector, which can be a problem at higher frequency because of limited common-mode rejection ratio of differential devices. *T-bridges* allow source and detector to share a common reference potential (usually the shield). Several configurations exist, most typical are the *bridged T* and the *twin-T* (Kugelstadt, 2002, Fig. 16–37), see Fig. 4.11.

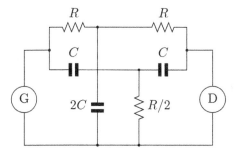

Figure 4.11
Twin-T bridge. With the particular values reported, the bridge is in equilibrium at $\omega = (RC)^{-1}$.

4.5   Transformer bridges

The use of transformers and inductive dividers (Sec. 3.3.2) in the realization of impedance bridges has first been considered by Blumlein (1928) and Walsh (1930). In ratio bridges, the advantages of an inductive divider with respect to an impedance divider are:

- the more convenient input and output impedances (input impedance is high for voltage dividers, low for current dividers; output impedance is low for voltage dividers, high for current dividers) which reduce loading effects;

- the small, and highly stable (in time, and respect to environmental conditions) ratio error.

The voltage and current ratio bridges of Fig. 4.10 can be realized with transformers, or inductive dividers, as shown in Fig. 4.12.

Consider the voltage ratio bridge of Fig. 4.12(a). The detector (an ammeter A) current reading I can be expressed as:

$$I = Y_2 V_2 - Y_1 V_1 = Y_2 (1 - k)V - Y_1 kV;$$

so at equilibrium ($I = 0$) one has

$$\frac{Y_2}{Y_1} = \frac{k}{1 - k}. \tag{4.6}$$

Similarly, consider the current ratio bridge of Fig. 4.12(b). The detector (a voltmeter V) voltage reading V is

$$V = Z_2 I_2 - Z_1 I_1 = Z_2 \frac{1}{1 - k} I - Z_1 \frac{1}{k} I;$$

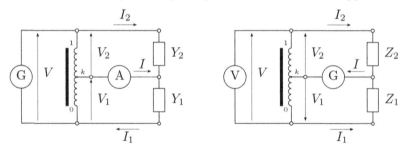

(a) Voltage ratio bridge with inductive voltage divider.

(b) Current ratio bridge with inductive current divider.

Figure 4.12
(a) Voltage and (b) current ratio bridges, corresponding to general bridges of Fig. 4.10, as realized with inductive dividers.

so at equilibrium $(V = 0)$ one has

$$\frac{Z_2}{Z_1} = \frac{1-k}{k},$$

which is the same equilibrium expression (4.6) of the voltage ratio bridge.

4.5.1 The current comparator

The detector V of the current ratio bridge of Fig. 4.12(b) can be substituted with a dedicated winding, which senses the magnetic flux Φ in the core. Such configuration is called a *current comparator* (Fig. 4.13) and is the most common current ratio bridge.

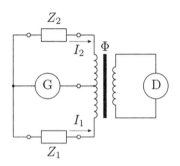

Figure 4.13
A current comparator.

The current comparator has some advantages over the voltage ratio bridge in particular situations:

high voltage impedance comparisons. The divider employed in the inductive voltage ratio bridge has to withstand the full voltage applied to impedances under comparison. High voltage at power frequency requires very large core sections and thick isolations. The current comparator divider works at low voltage, and can be a tabletop instrument even for high voltage measurements.

direct-current comparators. A unique feature of the current comparator is that it can be used with direct currents. The nonlinear properties of ferromagnetic core, typically annoying in instrument transformers, can be employed to detect dc flux unbalance by the *fluxgate* principle, Sec. 6.2.4.1. The resulting *direct-current comparator* (DCC) is the instrument of election for high-accuracy dc resistance and current measurements. See Moore and Miljanic (1988) for construction methods and extensive references.

cryogenic current comparators. The current comparator magnetic error is caused by flux leakage from the ferromagnetic core, which has a high but finite permeability. Flux leakage can be avoided by *superconducting shielding*, see Sec. 3.5. The resulting instrument (Harvey, 1972) is called *cryogenic current comparator* (CCC), since superconducting shielding operates at cryogenic temperatures. The cryogenic environment allows the use of one among the most sensitive magnetometric technique available, based on *superconducting quantum interference devices* (SQUIDs) which employ macroscopic quantum charge transport effect to achieve extreme sensitivity (down to $fT\,Hz^{-\frac{1}{2}}$) to both dc and low-frequency magnetic induction.

Direct-current CCC, although research instruments, are now widespread in metrology laboratories; automated bridges (Williams et al., 2010) can achieve relative accuracies of parts in 10^{10}.

A combination of superconducting shielding and coaxial winding has been employed to construct an ac CCC insensitive to both magnetic *and* capacitive errors by Grohmann and Hechtfischer (1984). The technique at the moment has not been developed further, but has not been criticized in literature. Therefore, the potential of an ac CCC for ultimate-accuracy impedance comparison remains unexplored to date.

4.5.2 Equal power transformer ratio bridge

The comparison of resistive impedances R_1 and R_2 having nominal ratio $R_1/R_2 = q$ with the voltage or current ratio transformer bridges of Fig. 4.12 results in different power dissipation P_1 and P_2 from resistors R_1 and R_2; in a voltage ratio bridge $P_1/P_2 = q$, whereas in a current ratio bridge

$P_1/P_2 = q^{-1}$. The power ratio is the electrical dissipation of resistors R_1 and R_2 under comparison. Since the maximum measurement power is determined by the standards' specifications (often $P = 1\,\text{mW}$ for medium-valued resistance standards), for large q one of the standards is excited with too low electrical power, and bridge sensitivity is compromised.

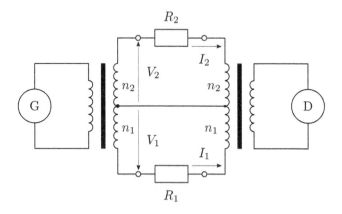

Figure 4.14
A 2T equal power ratio bridge, having ratio $(n_1/n_2)^2$.

The equal-power transformer ratio bridge (Blumlein, 1946; Hall, 1970), Fig. 4.14, is a combination of a voltage ratio and a current ratio bridge. If both transformer/dividers have the same secondary turns ratio n_1/n_2, the bridge ratio is $q = (n_1/n_2)^2$, and the same power $P_1 = P_2$ is dissipated in both standards. The typical arrangement for resistance scaling involves $q = (10)^2 = 100$.

4.6 Beyond 2T definition of impedance standards

As said in Ch. 2, the 2T definition of impedance is often inadequate. This raises the problem of how to realize voltage/current bridges, which allow for more complex impedance definitions. We provide some examples of how this can be achieved. Most examples deal with voltage ratio bridges, since this is the most common condition encountered in real setups; however, dual current bridges can be conceived.

4.6.1 3T, 2P

3T and 2P definitions have different topologies, but a similar defining condition (see Ch. 2): for 3T a null low-to-shield potential difference, for 2P a null low-voltage port potential difference.

Fig. 4.15 shows the transformation of a 2T voltage ratio bridge (a) in the corresponding 3T (b) or 2P bridges (c). The 2T topology is expanded to 3T by adding electrostatic shields, or to 2P by adding coaxial outer conductors. Fig. 4.16 shows a 2P bridge employing an inductive voltage divider as a ratio arm, to achieve a comparison bridge for impedances having identical phase angle.

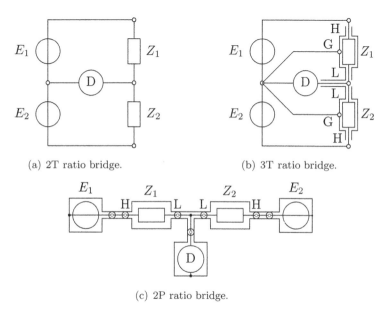

(a) 2T ratio bridge.　　　　　(b) 3T ratio bridge.

(c) 2P ratio bridge.

Figure 4.15
The transformation of a 2T (a) ratio bridge to the corresponding 3T (b) and 2P (c) ratio bridges.

4.6.2 Wagner balance

In the bridges of 4.15(b) and 4.15(c), the shunt admittances between inner conductors and shielding are energized by the very same generators that achieve the reference voltage ratio. Real-world ratio arms, however, are sensitive to current loading, which can alter the reference ratio.

Consider as an example the 2P transformer bridge of Fig. 4.16. The capacitive shunt admittances between inner and outer conductors of the coaxial

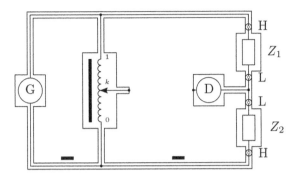

Figure 4.16
2P ratio bridge based on an inductive voltage divider.

mesh, if not properly balanced, give rise to a net current flowing in the shorted tap of the IVD, causing a ratio error.

A solution is given by the use of *Wagner balance*, or *Wagner ground* (Wagner, 1911). To the original bridge (Fig. 4.16), an additional ratio arm, the *Wagner arm*, is added (Fig. 4.17). The source, the Wagner arm, and stray admittances of the coaxial mesh constitute an additional impedance bridge. Such a bridge can be balanced by adjusting the components in the Wagner arm to null the potential difference between a chosen point in the inner conductor network, and the corresponding point on the shield network. In the bridge of Fig. 4.17, the point of interest is the IVD tap; to trim the Wagner arm, the short is temporarily substituted with a detector, the balancing is performed, and the short reinserted (this last move prevents errors caused by small drifts in the Wagner balance to cause reading errors, and reduces the bridge noise).

The Wagner ground can be implemented in various ways; Wolff (1959) discuss a number of possible Wagner networks. A Wagner balance can be achieved also with an active generator, either a voltage generator driving the shield to the desired potential (Corney, 1979, 2003) or a current generator supplying the required current imbalance.

4.6.3   4T

4T definition of impedance can be readily achieved with in implementations of the *I-V* method (Fig. 4.1) or two-voltage methods (Fig. 4.3). The realization of a 4T ratio bridge is less obvious: voltage drops V_1 and V_2 on impedances Z_1 and Z_2 cannot be directly compared with the three-terminal bridge ratio arms (either built with by impedances, or by an IVD) considered until now.

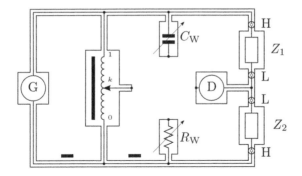

Figure 4.17
The 2P IVD ratio bridge of Fig. 4.16, provided with Wagner balance arm
elements C_W and R_W. If the Wagner balance process asks for negative values
for C_W or R_W, these are simply connected to the other side of the bridge.

4.6.3.1 Anderson loop

The so-called *Anderson loop* (Anderson, 1994, 1998) uses a double-differential
linear electronic amplifier to obtain the difference $V_1 - V_2$. The measurement
accuracy is limited by the gain accuracy and common-mode rejection ratio
(CMRR) of the amplifier. In instrumentation amplifiers, CMRR falls down
with frequency, resulting in the main limitation in accuracy of the Anderson
loop for impedance measurements.

4.6.3.2 Difference transformer

The (weighted) difference can be achieved with a sum-difference transformer;
the resulting bridge has the same schematics of Fig. 4.8(b), V_m being replaced
with a detector.

4.6.3.3 Active compensation

The voltage drop zI caused by the impedance z of the connection between
impedance current terminals can be compensated by the injection of a voltage
V_{G12}. Fig. 4.18 shows a 4T bridge with active compensations.[2]

V_{G12} is typically injected in the current mesh with a feedthrough transformer. The synthesis of V_{G12} can be achieved with a passive network,
with linear electronics (Vyroubal, 1993), or with more advanced modula-
tion/demodulation techniques (Cabiati and La Paglia, 1977; Callegaro and
D'Elia, 1998).

---

[2]Forms of passive compensations, where the voltage drop between current terminals is
reinjected in the voltage mesh with a transformer, have also been experimented (Foord
et al., 1963).

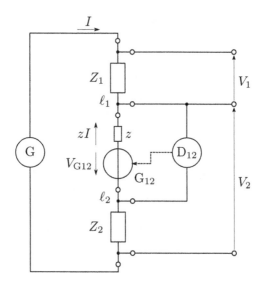

Figure 4.18
A 4T bridge (ratio arm is omitted) with active compensation. Detector D_{12}
senses the voltage drop zI on the current path between nodes ℓ_1 and ℓ_2. A voltage generator G_{12} injects a compensating voltage V_{G12}, which is trimmed until
D_{12} reads zero, so $V_{G12} = zI$. Now voltage drops V_1 and V_2 on impedances
under comparison Z_1 and Z_2 are referenced to a single mesh node and a three-
terminal ratio arm (two reference impedances, or an IVD) can be employed
for the determination of V_1/V_2.

4.6.3.4 The Kelvin bridge

The Kelvin double bridge, shown in Fig. 4.19(a), is a generalization of the Wheatstone bridge suitable for the comparison of two low-valued impedances Z_1 and Z_2, defined as four-terminal standards. The (unknown) impedance of the current path between inner nodes ℓ_1 and ℓ_2 of Z_1 and Z_2 is modeled by a lumped impedance z. The impedances z_1 and z_2 give the so-called *Kelvin arm*.

The bridge can be analyzed by applying a delta-wye transformation of impedances z_1, z_2, and z into z_A, z_B, and z_C as in Fig. 4.19(b). The transformation equations for z_A and z_B are

$$z_A = \frac{z\,z_1}{z + z_1 + z_2}; \qquad z_B = \frac{z\,z_2}{z + z_1 + z_2}, \tag{4.7}$$

The transformation maps the original network into a Wheatstone bridge, where impedances under comparisons are $Z_1 + z_A$, $Z_2 + z_B$, Z_3 and Z_4. Impedance z_C is in series with D and thus modifies its sensitivity but not the equilibrium equation (4.5) of the Wheatstone bridge, which can be written

$$\frac{Z_1 + z_A}{Z_2 + z_B} = \frac{Z_3}{Z_4}. \tag{4.8}$$

After the delta-wye substitution (4.7), Eq. (4.8) can be rewritten as

$$\frac{Z_1\,(z + z_1 + z_2) + z\,z_1}{Z_2\,(z + z_1 + z_2) + z\,z_2} = \frac{Z_3}{Z_4}. \tag{4.9}$$

The balance equation (4.9) is independent of the value of z if

$$\frac{z_1}{z_2} = \frac{Z_1}{Z_2} = \frac{Z_3}{Z_4},$$

called the *Kelvin condition*. It can be achieved by tracking (by adjusting the nominal values of the variable components z_1 and z_2), as it happened in some instruments of the past (with a special arrangement of dial switches).

The Kelvin condition can be achieved also by a direct alteration of the voltage drop zI, either by a variation of z (for example, by disconnection, so $z \to \infty$; or, less brutally, by introducing an additional series resistor), or by adding a small voltage (e.g., with a feedthrough transformer, Sec. 3.3.9). The condition is achieved by adjusting z_1 and z_2 in a way such that the equilibrium condition becomes insensitive to the variation of zI.

The Kelvin bridge can be generalized to the Warshawsky (1955) bridge, where all the four impedances of the original Wheatstone bridge are defined as four-terminal standards.

4.6.3.5 Combining network realization

The Kelvin arm is an example of a *combining network*: impedances z_1 and z_2 achieve at node ℓ an adjustable linear combination of the potentials at

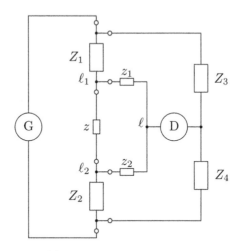

(a) Kelvin bridge schematics.

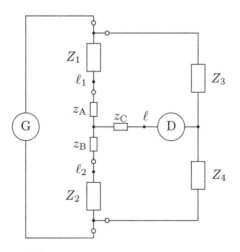

(b) After delta-wye transformation.

Figure 4.19
The Kelvin double bridge. (a) bridge schematics. z models the impedance of the current path between nodes ℓ_1 and ℓ_2. (b) After the delta-wye transformation of $\{z_1, z_2, z\}$ in $\{z_A, z_B, z_C\}$.

nodes ℓ_1 and ℓ_2; the combining network is adjusted to have the potential in ℓ independent of the unknown voltage drop zI.

Inductive voltage dividers permit the realization of practical and stable combining networks, see Fig. 4.20.

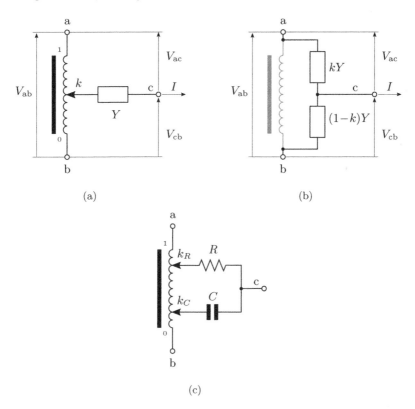

(a)   (b)

(c)

Figure 4.20
(a) A combining network realized with a variable IVD and an admittance Y. For an ideal IVD, having zero input admittance and output impedance, network (a) is equivalent to network (b), which shows more clearly the behavior as a Kelvin arm (the IVD is grayed because it becomes immaterial). (c) General combining network composed of a double-output IVD and two admittances in quadrature (a resistor R and a capacitor C); the network permits an adjustment in magnitude and phase of the impedance ratio given in the equivalent circuit (b).

Several 4T ratio bridges based on inductive ratio and Kelvin arms have been published (Gibbings, 1962; Moore and Basu, 1966; Hanke, 1978). An example is shown in Fig. 4.21.

In some cases a combining network can produce systematic errors (Melcher,

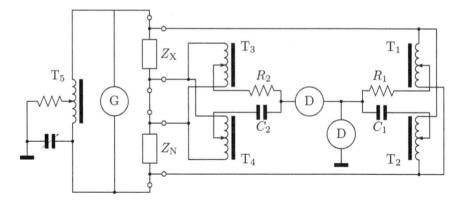

Figure 4.21
A Kelvin double bridge with inductive ratio arms. T_1 and T_2 are the main ratio arm dividers, T_3 and T_4 constitute the Kelvin ratio arm, T_5 the Wagner arm. Z_N and Z_X are the impedances under comparison. Adapted from Hanke (1978).

1994). Combining networks and active compensation can be integrated for improved performance (Hanke and Ramm, 1983).

4.6.4   4P

4P impedance definition is the most accurate, and ask for the most complex additional circuitry to be achieved. Several approaches can be followed, and also combined together to obtain a 4P bridge.

4.6.4.1   Correction method

2P bridges can be employed in the measurement of 4P impedances with the *extrapolation* method (Shields, 1974), where 2P measurements are numerically corrected to match the corresponding 4P definition. The corrections are derived from additional 2P measurements of elements of the 4×4 impedance matrix of the standard, and by artificially enhancing the effect of shunt admittances of cables and extrapolate the effect to zero.

4.6.4.2   Compensation

4P *I-V* measurement is implemented in LCR meters, see Fig. 4.28. A more complete compensation circuit is shown in Fig. 4.22, which can be thought of as a conversion circuit from a 4P standard to a 2P one having the same transadmittance Y_{4P}. Such compensation scheme can be applied to any bridge

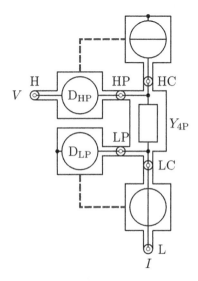

Figure 4.22
A 4P compensation scheme that converts the 4P passive admittance standard Y_{4P} in a 2P one. If detectors D_{HP} and D_{LP} are set to zero by acting on the current and voltage compensation generators, voltage V applied at port H gives a current $I = Y_{4P}V$ at port L.

schematics to obtain a 4P bridge: Fig. 4.23 shows a milestone example given by Cutkosky (1970).

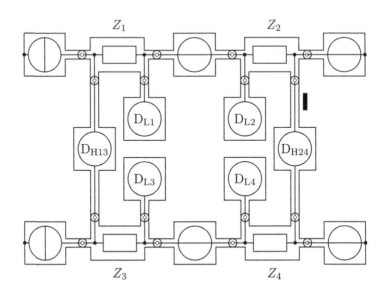

Figure 4.23
Principle schematics of a 4P Wheatstone bridge. The six generators are trimmed to null all six network detectors. Adapted from Cutkosky (1970, Fig. 2).

Such a direct compensation scheme asks for highly sensitive detectors and stable generators, since any drift from perfect null condition causes measurement errors. In the past (Cutkosky, 1970), it has been considered only as a didactical principle. The evolution of digital bridges, Ch. 5, however, permits a direct implementation of the scheme.

4.6.4.3   Combining network

The combining network principle can be employed in coaxial bridges to achieve 4P definition. Fig. 4.24 shows a voltage ratio arm constructed with two 4P impedances Z_1 and Z_2, which can be embedded in a 4P voltage ratio bridge, as in Fig. 4.25.

Kibble and Rayner (1984) and Awan et al. (2011) are devoted to the explanation of principles, detailed schematics, and operating procedures of 4P bridges for primary impedance measurements, where the combining network principle is extensively employed. A 4P Maxwell-Wien bridge for inductance calibration has also been reported (Côté, 2009).

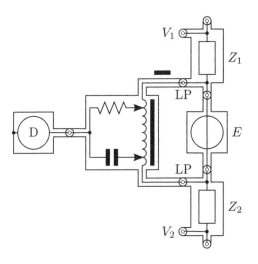

Figure 4.24
Series connection of two 4P impedance standards Z_1 and Z_2, with the help of a combining network and an adjustable voltage source E (which can be given by the voltage drop on a variable impedance, or a small voltage derived from main source and injected with a feedthrough transformer). The network and the external circuitry are trimmed until the null reading of D is independent of a variation of E. The resulting series can be employed as voltage ratio arm in a coaxial bridge: under the equilibrium condition described, the relation $V_1/V_2 = Z_1/Z_2$ holds.

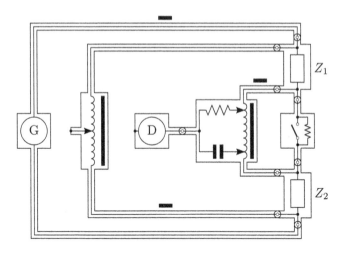

Figure 4.25
The 4P ratio arm of Fig. 4.24 embedded in the IVD ratio bridge of Fig. 4.16 to achieve a 4P IVD ratio bridge (quadrature compensation, and Wagner arm, are omitted).

4.6.5 The quadrature bridge

Let us consider again the voltage ratio bridge of Fig. 4.10(a). The instrument accuracy is limited by the accuracy of the amplitude ratio and phase difference of generators E_1 and E_2. Inductive voltage dividers permit to achieve high accuracy amplitude ratios, but only when the relative phase difference is either 0 or π, useful when comparing *like* impedances, such as in $R - R$ or $C - C$ comparisons. When dealing with a $R - C$ comparison, as in Fig. 4.26(a), E_1 and E_2 must have a phase difference near to $\pm\pi/2$.

The quadrature bridge is a *double bridge*.[3] The voltage ratio bridge structure of Fig. 4.26(a), measuring standards R and C can be symmetrically duplicated, Fig. 4.26(b), by adding a third voltage source E_1^* and two additional standards R^* and C^*. For pure impedances, the double equilibrium is achieved when

$$j\omega C\, E_1 + \frac{1}{R}\, E_2 = 0,$$

$$j\omega C^*\, E_2 + \frac{1}{R}\, E_1^* = 0,$$

---

[3] Millea and Ilie (1969) gives an approach more general than the present one to double bridges and quadrature bridges, including RL and LC comparison circuits.

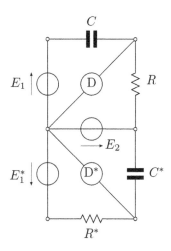

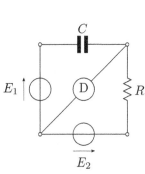

(a) The voltage ratio bridge of Fig. 4.10(a), when performing a RC comparison and redrawn to show the phase shift between E_1 and E_2.

(b) The quadrature bridge.

Figure 4.26
(a) Single RC bridge; (b) quadrature bridge, obtained by doubling the RC bridge.

which, by eliminating E_2, reduce to

$$\omega^2 R R^* C C^* E_1 + E_1^* = 0;$$

If $E_1^* = -E_1$ (which can be obtained with a 1:1 inductive voltage divider), the quadrature bridge balance equation

$$\omega^2 R R^* C C^* = 1 \tag{4.10}$$

is obtained. Eq. (4.10) does not contain the ratio E_2/E_1, which is difficult to realize or measure accurately in the single bridge of Fig. 4.26(a).

Typically, the bridge is operated with four standards having equal nominal impedance at the working frequency, $R = R^* = \omega C = \omega C^*$; for such cases, $E_2 = -\mathrm{j}E_1$. However, so-called *multi-frequency* quadrature bridges introduce additional voltage ratio elements (inductive voltage dividers) to relax the condition of having equal impedance moduli, thus permitting the measurement of the same RC value combinations at different frequencies (Nakamura et al., 1999; Bohacek, 2004).

As drawn in Fig. 4.26(b), the quadrature bridge is still very sensitive to the stability of E_1/E_2, since this ratio influence both detectors. The use of a combining network permits to create a detector insensitive to E_2, shown in Fig.

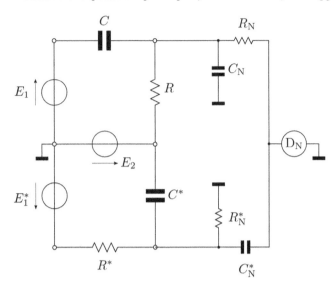

Figure 4.27
Providing the quadrature bridge of Fig. 4.26(b) with a combining network, composed by elements R_N, R_N^*, C_N, C_N^*, can provide a detector point D_N insensitive to E_2.

4.27. The network is given by elements R_N, R_N^*, C_N, C_N^*, with $R_N \approx R_N^* \approx R$ and $C_N \approx C_N^* \approx C$; C_N and R_N^* are adjustable elements.

The combining network is trimmed by setting $E_1 = E_1^* = 0$, so R is in parallel with R_N and C^* with C_N^*. In this condition, the resulting configuration is that of a twin-T bridge (Sec. 4.4.2), having E_2 at input and D_N as output, which can be balanced at frequency ω by acting on the adjustable components of the combining network. Since the quadrature bridge network is linear, the combining network continue to reject E_2 also when generators E_1 and E_1^* are switched on again.

4.7　The LCR meter

The *LCR* (or *RLC*) *meter*, or *bridge*, is an electronic instrument specifically designed for the measurement of impedances having arbitrary phase angle (LCR stands for inductance, capacitance, resistance); see Hall (2004) for an history of its development.

Impedance standards are connected to the meter with cables; components with fixtures (Sec. 2.3). The impedance definition employed by the bridge is dependent on its accuracy class; more accurate bridges employ 4P definition.

The meter includes an adjustable (in frequency and amplitude) sinewave generator to energize the device under measurement; readings are couples of values, displayed according to the selected representation (Sec. 1.9). The meter includes one or more impedance standards, not accessible by the user, and programmable-gain amplifiers. Standards and amplifier gains are internally switched according to the chosen *range*; the number of ranges can vary from a few to more than a hundred. The autoranging mode is always available and usually set as default. Programming and data processing function, such as frequency sweeps, graphical representations, pass/fail tests are often implemented in the instrument firmware.

A principle schematics of LCR bridge employing 4P impedance definition is shown in Fig. 4.28.

The main generator G energizes impedance Z through its high current port HC; it can be adjusted by the user to desired frequency and amplitude (voltage or current), and dc bias.

Vector voltmeter V is connected to the high voltage port HP, and measures V_{HP}. Vector ammeter A is connected to low current port LC and measures I_{LC}.

Detector D monitors the 4P definition condition $V_{LP} \equiv 0$; an auxiliary voltage generator G' is continuously adjusted by an automated control to null D. Either linear control or more sophisticated techniques, such as synchronous filtering, can be employed.

Internal software computes $Z = V_{HP}/I_{LC}$, which is displayed in the representation (Sec. 1.4.3) chosen by the user.

Real instruments can implement a two-voltage method, Sec. 4.2. A is realized with a current-to-voltage converter (Sec. 3.2.2), and a single vector voltmeter is switched between V and A positions.

4.7.1 Ranges

Frequency ranges available on LCR meters vary considerabily. Portable 2T instruments often have a single working frequency, others provide a few centered in the audio range (for example, on Agilent 4263B 100 Hz, 120 Hz, 1 kHz, 10 kHz, 20 kHz, 100 kHz are available); high-accuracy instruments provide a wider frequency range (typically 20 Hz to a few MHz) with good frequency resolution (e.g., 5-digit for Quadtech mod. 7600), with some instruments extending the range to 100 MHz and beyond (e.g., Agilent 4294A, Wayne-Kerr 6500P).

Amplitude range of the test signal typically spans from around 10 mV to 10 V, with currents typically limited to 100 mA maximum. However, external sets or options can expand these ranges and add dc bias voltage and high-current capabilities, which can be of interest in material and device testing.

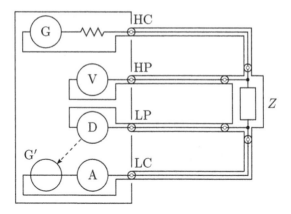

Figure 4.28
Principle schematics of a 4P LCR bridge.

Amplitude range is usually frequency-dependent, with more setting limitations for higher frequencies.

4.7.2　Accuracy

The accuracy specifications of an LCR meter are strongly dependent on the impedance magnitude, on amplitude and frequency of the test signal, on the chosen representation for reading, on the chosen measurement speed. The accuracy is given by a combining expression, whose terms are available as tables or charts. Best relative accuracy claims are in the 10^{-4} range, but can drop to 100 % at impedance or excitation range extremes, or when an unfortunate representation is chosen (e.g., when performing measurements on high-Q standards).

Resolution, stability, and linearity of the LCR meter can be much better than the accuracy in the corresponding measurement condition. Top-class instruments have 6-digit displays, and the linearity can be within a few parts in 10^6 (Suzuki et al., 2001; Mitsuo et al., 2007).

4.7.3　The LCR meter as a comparator

It is possible to employ LCR bridges to perform impedance measurements with improved accuracy with respect to the specifications given by the manufacturer for a direct measurement.

4.7.3.1 Substitution measurements

Substitution measurements can be performed easily with a LCR bridge. The impedance Z_X to be measured is compared with a calibrated standard Z_S having the same nominal value. The LCR meter is connected in sequence to the two standards, giving readings Z_S^m and Z_X^m. Two measurement model equations ("difference" and "ratio") can be applied:

$$Z_X = (Z_X^m - Z_S^m) + Z_S; \tag{4.11}$$

$$Z_X = \frac{Z_X^m}{Z_S^m} \cdot Z_S. \tag{4.12}$$

The rationale of Eq. (4.11) is the correction of the LCR meter offset near the working point, whereas Eq. (4.12) is a correction of its gain. Since most measurements with LCR meters are performed after open-short calibration of the same (Sec. 4.8.2.3), offset is already corrected, hence Eq. (4.12) may be considered more appropriate (Callegaro, 2001).

To improve the precision of the method, several measurement cycles can be performed, and averages taken. Automation is possible but require high-performance coaxial switches with low crosstalks (Callegaro et al., 2002) not usually available on the market.

4.7.3.2 Scaling measurements

Aoki and Yokoi (1997) introduced an extension of the substitution method. Two impedances Z_1 and Z_2 of the same kind but having nominal values in a ratio $k = Z_2/Z_1$ are involved (e.g., two capacitors having 1:10 nominal capacitance ratio). The method is shown in Fig. 4.29. Two different measurement configurations are employed to measure Z_1 and Z_2; the rationale is to feed the LCR meter with similar voltages and currents in both configurations, in order to achieve similar readings, like in substitution measurements.

Let's assume $k < 1$, i.e. $|Z_2| < |Z_1|$. For the measurement of the lower-value impedance Z_2, Fig. 4.29(b), the bridge is employed directly. In the measurement of the higher-valued impedance Z_1, Fig. 4.29(a), the generator (connected to port 1 of the standard) has to be amplified by a factor $1/k$ and the voltage at port 2 is fed to a calibrated inductive voltage divider set at ratio k.

The method has been developed to calibrate high-value ($10\,\text{nF} - 1\,\mu\text{F}$) solid-dielectric capacitors, by successive 10:1 scaling, using a $1\,\text{nF}$ gas-dielectric capacitor as reference standard, at frequencies up to $100\,\text{kHz}$ (Aoki and Yokoi, 1997; Avramov-Zamurovic et al., 2007).

A dual method can be conceived, by using a current ratio divider.

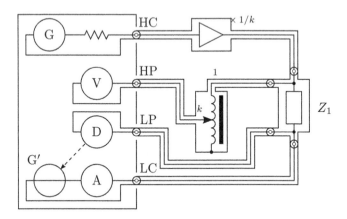

(a) Measurement of the high-valued impedance Z_1.

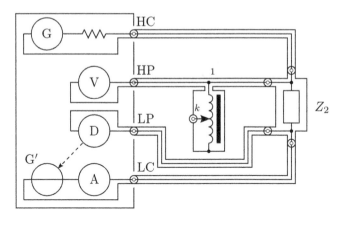

(b) Measurement of the low-valued impedance Z_2.

Figure 4.29

Scaling measurement by substitution with a LCR meter, of two impedances having nominal ratio $Z_2 = kZ_1$ (with $k < 1$). Adapted from Aoki and Yokoi (1997).

4.7.4 The LCR meter as a transformer bridge

The LCR meter circuitry can be employed with a transformer or inductive divider (3.3.2) to act as a voltage, or current, ratio transformer bridge (Kim and Semenov, 2008), see Fig. 4.30.

4.8 Calibration

The *calibration* of an impedance meter is the operation that establishes a relation, typically called a *correction*, between the (known) impedance values of appropriate set of impedance standards and the corresponding meter indications (readings); in a second step, this relation is employed to provide a measurement result from the indications obtained, with the same instrument, on impedances under measurement (VIM, 2.39).

Modern instruments, such as network analyzers or LCR meters, have functions to automate the calibration process and apply correction to readings; calibration is in this case an improper term, since one should talk of *adjustment* (VIM, 3.11).

The term *compensation* may refer to a particular calibration, the main purpose of which is to take into account the effect of the connections and fixtures employed in the measurement on the meter reading. In literature the terms *calibration, compensation, correction*, and *adjustment* are often employed indifferently.

Calibration schemes are designed by modeling the relevant measurement errors caused by internal imperfections of the instrument and of the connection, called *error terms*. After the model is achieved, the calibration procedure has to involve a sufficient number of impedance standards, and corresponding meter readings, in order to solve the model and identify the error terms, which are stored in memory. Further readings will be corrected with the model employing the error terms thus identified. For example, LCR meter substitution measurement, Sec. 4.7.3.1, can be considered a simple meter calibration, employing only one impedance standard. Model (4.11) considers and corrects only the offset error of the meter; model (4.12) only the gain error.

Under meter linearity hypothesis, several error models commonly employed can be expressed as a linear fractional transformation, also called *Möbius transformation*,[4] of the impedance value Z into the meter reading Z^{m}:

$$Z^{\mathrm{m}} = f(Z) = \frac{AZ + B}{CZ + D}, \qquad AD - BC \neq 0. \qquad (4.13)$$

The generic Möbius transformation is equivalent to a sequence of simpler transformations of translation (which can model offset errors), inversion and

---

[4]Special Möbius transformations have been already introduced in Sec. 1.9.

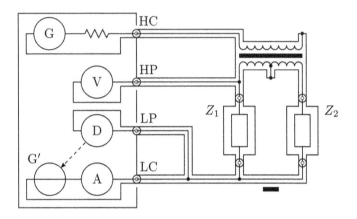

(a) Voltage ratio bridge

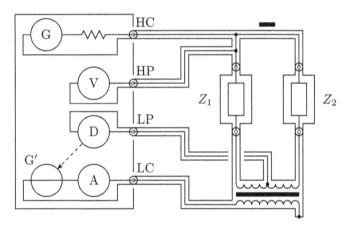

(b) Current ratio bridge

Figure 4.30
LCR meter employed, with a transformer, as a voltage (a) or current (b) ratio transformer bridge for the comparison of 2P impedances Z_1 and Z_2. Adapted from Kim and Semenov (2008).

reflection, dilation, and rotation (which can model phase shifts and gain errors) in the Z complex plane. After the calibration process, if the coefficients A, B, C, D are properly identified, a correction can be applied to Z^m with the inverse transformation, which is a Möbius transformation also:

$$Z = f^{-1}(Z^m) = \frac{D\,Z^m - B}{-C\,Z^m + A}. \tag{4.14}$$

4.8.1   Compensation

Compensation of meter connections gives a physical example of the occurence of Eq. (4.13). Consider Fig. 4.31, where a meter is connected to the 2T impedance Z.

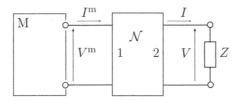

Figure 4.31
2T compensation model; network $\mathcal{N}$ models the connection from meter M to impedance Z.

The electromagnetic effects of the connection (including cables, connectors, fixtures) can be modeled with the two-port passive network $\mathcal{N}$. Voltage V and current I present on impedance Z are related to meter readings, V^m and I^m, by the $ABCD$ transmission matrix (Sec. 1.6.3.4) of $\mathcal{N}$:

$$\begin{bmatrix} V^m \\ I^m \end{bmatrix} = \begin{bmatrix} A & B \\ C & D \end{bmatrix} \cdot \begin{bmatrix} V \\ I \end{bmatrix}; \tag{4.15}$$

for example, if the connection is a transmission line, Sec. 1.6.4.4, the $ABCD$ matrix is given by Eq. (1.13).

The meter displays the reading Z^m, which can be written by Eq. (4.15) as

$$Z^m = \frac{V^m}{I^m} = \frac{A\,V + B\,I}{C\,V + D\,I} = \frac{A\,Z + B}{C\,Z + D}. \tag{4.16}$$

4.8.2   Calibration schemes

Without loss of generality, Eqs. (4.13) and (4.14) can be rewritten as

$$Z^m = \frac{\alpha\,Z + \beta}{\gamma\,Z + 1} \tag{4.17}$$

and

$$Z = \frac{Z^{\mathrm{m}} - \beta}{-\gamma Z^{\mathrm{m}} + \alpha}. \tag{4.18}$$

The three coefficients α, β, and γ can be determined with the measurement of at least three impedance standards of known value Z_1, Z_2, and Z_3. In sequence, Z_k ($k = \{1, 2, 3\}$) is connected to the impedance meter, and a reading Z_k^{m} is taken. Z_k and Z_k^{m} are related by Eq. (4.17), which gives a system of three equations in unknowns α, β, and γ:

$$\{_{k=\{1,2,3\}}} \quad Z_k\,\alpha + \beta - Z_k\,Z_k^{\mathrm{m}}\,\gamma = Z_k^{\mathrm{m}}\,. \tag{4.19}$$

4.8.2.1   Three-standard calibration

Three-standard calibration scheme relies on the general solution of system (4.19). The solution, combined with Eq. (4.18), gives for the generic impedance Z_{X} (if $Z_{\mathrm{X}}^{\mathrm{m}}$ is the corresponding meter reading):

$$
\begin{aligned}
Z_{\mathrm{X}} = &\frac{N}{D}, \qquad \text{where} \\
N = &\, Z_1 Z_2 Z_2^{\mathrm{m}} Z_3^{\mathrm{m}} - Z_1 Z_2 Z_1^{\mathrm{m}} Z_3^{\mathrm{m}} + Z_1 Z_2 Z_1^{\mathrm{m}} Z_{\mathrm{X}}^{\mathrm{m}} - Z_1 Z_2 Z_2^{\mathrm{m}} Z_{\mathrm{X}}^{\mathrm{m}} \\
&+ Z_1 Z_3 Z_1^{\mathrm{m}} Z_2^{\mathrm{m}} - Z_1 Z_3 Z_2^{\mathrm{m}} Z_3^{\mathrm{m}} - Z_1 Z_3 Z_1^{\mathrm{m}} Z_{\mathrm{X}}^{\mathrm{m}} + Z_1 Z_3 Z_3^{\mathrm{m}} Z_{\mathrm{X}}^{\mathrm{m}} \\
&- Z_2 Z_3 Z_1^{\mathrm{m}} Z_2^{\mathrm{m}} + Z_2 Z_3 Z_1^{\mathrm{m}} Z_3^{\mathrm{m}} + Z_2 Z_3 Z_2^{\mathrm{m}} Z_{\mathrm{X}}^{\mathrm{m}} - Z_2 Z_3 Z_3^{\mathrm{m}} Z_{\mathrm{X}}^{\mathrm{m}}, \\
D = &\, Z_1 Z_1^{\mathrm{m}} Z_2^{\mathrm{m}} - Z_1 Z_1^{\mathrm{m}} Z_3^{\mathrm{m}} - Z_1 Z_2^{\mathrm{m}} Z_{\mathrm{X}}^{\mathrm{m}} + Z_1 Z_3^{\mathrm{m}} Z_{\mathrm{X}}^{\mathrm{m}} \\
&- Z_2 Z_1^{\mathrm{m}} Z_2^{\mathrm{m}} + Z_2 Z_2^{\mathrm{m}} Z_3^{\mathrm{m}} + Z_2 Z_1^{\mathrm{m}} Z_{\mathrm{X}}^{\mathrm{m}} - Z_2 Z_3^{\mathrm{m}} Z_{\mathrm{X}}^{\mathrm{m}} \\
&+ Z_3 Z_1^{\mathrm{m}} Z_3^{\mathrm{m}} - Z_3 Z_2^{\mathrm{m}} Z_3^{\mathrm{m}} - Z_3 Z_1^{\mathrm{m}} Z_{\mathrm{X}}^{\mathrm{m}} + Z_3 Z_2^{\mathrm{m}} Z_{\mathrm{X}}^{\mathrm{m}}.
\end{aligned} \tag{4.20}
$$

If more than three Z_k standards are employed, system (4.19) becomes overdetermined and has to be solved in the context of linear regression (e.g., by least-squares method) or by making suitable assumptions on standard properties; an example is a *two-load* calibration, actually employing four impedance standards (Yonekura, 1994; Yonekura and Jansons, 1994).

4.8.2.2   SOL calibration

For the limiting case, where $Z_{\mathrm{O}} \equiv Z_1 \to \infty$ (an *open*), $Z_{\mathrm{S}} \equiv Z_2 = 0$ (a *short*), and $Z_{\mathrm{L}} \equiv Z_3$ (a *load*), Eq. (4.20) simplifies to (Agilent 346-3):

$$Z_{\mathrm{X}} = \frac{(Z_{\mathrm{L}}^{\mathrm{m}} - Z_{\mathrm{O}}^{\mathrm{m}}) \cdot (Z_{\mathrm{S}}^{\mathrm{m}} - Z_{\mathrm{X}}^{\mathrm{m}})}{(Z_{\mathrm{L}}^{\mathrm{m}} - Z_{\mathrm{S}}^{\mathrm{m}}) \cdot (Z_{\mathrm{O}}^{\mathrm{m}} - Z_{\mathrm{X}}^{\mathrm{m}})} \cdot Z_{\mathrm{L}}, \tag{4.21}$$

which is called *short-open-load* calibration, or calibration, equation.

It has been shown (Suzuki et al., 1993) that Eq. (4.21) is valid also for a 4P meter and connections. Since 4P impedance definition permits the realization of open and short standards having extremely small stray parameters (Sec.

8.7), hence approaching the ideal behavior $|Z_O| = \infty$, $Z_S = 0$, SOL calibration is typically implemented in 4P LCR meter firmware. When SOL calibration is applied to network analysis, Z_L is the reference impedance Z_0.

A special application of SOL calibration has been proposed for the measurement of 4P resistance standards with reference capacitors, at frequencies up to 1 MHz (Svetik and Lapuh, 2000; Suzuki et al., 1993).

4.8.2.3 Open-short calibration

Open-short calibration make the assumption that $\alpha = 1$ (no gain error) in Eq. (4.18), and therefore of two unknowns in system (4.19). Further, perfect open and short standards are employed. This gives $\beta = Z_S^m$, $\gamma = (Z_O^m)^{-1}$, and the resulting open-short calibration equation is

$$Z_X = \frac{Z_X^m - Z_S^m}{1 - \dfrac{Z_X^m}{Z_O^m}}.$$

4.9 Resonance methods

Resonance methods are employed for the measurement of reactive impedances: inductors, or capacitors.

The reactive impedance can be connected in series or parallel with a known reactance of the opposite sign, to create a LC resonator of impedance $Z_{LC}(\omega)$. Resonance can be detected by measuring $|Z_{LC}(\omega)|$ versus frequency ω and by looking at its minimum (*series* resonance) or maximum (*parallel* resonance); the detection circuit does not require particular accuracy, or to be sensitive to phase.

The resonant circuit can be embedded in an electronic oscillator, which thus gives a signal of frequency ω as output, resulting in an impedance-to-frequency converter. If the resonance has an high Q, the measurement can achieve extremely high sensitivity.

Real inductors and capacitors display *self-resonances* with the stray parameters of the construction (on inductors, the capacitance between windings and to the shield; on capacitors, the self-inductance of electrodes and connections).

4.9.1 Inductance

The measurement of self-inductance L with the resonance method requires a variable capacitor C as a reference standard. A series or parallel LC circuit is created, and C is tuned until the LC circuit resonates and thus behaves as a pure resistor. Fig. 4.32 shows the equivalent circuits of a lossy inductor $L_S - R_S$ (or $L_P - R_P$) in series (or parallel) with a lossless variable capacitor C. At resonance, $\omega^2 L_{S,P} C = 1$.

To sense the resonance condition, any noncalibrated measurement of the impedance (admittance) magnitude $|Z(\omega)|$ (or $|Y(\omega)|$), which reaches an extremum in the resonance condition, can be employed.

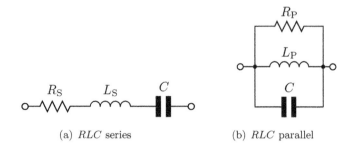

(a) *RLC* series   (b) *RLC* parallel

Figure 4.32
(a) Series and (b) parallel *RLC* equivalent circuits of a lossy inductor in parallel/series with a capacitor. In the resonance condition, these combinations are called *resonant branches*.

The sensitivity of the method is dependent on the inductor Q. For this reason, if the resonance method is employed for a primary realization of inductance unit, cylindrical inductors are preferred over toroidal ones (Rayner et al., 1980).

In the past, a popular instrument called the *Q-meter*, based on the resonance method, was employed for a number of different impedance measurements on radio circuit components. The basic circuit is shown in Fig. 4.33.

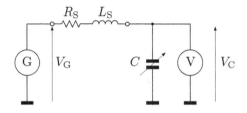

Figure 4.33
Q-meter basic circuit. At resonance, $\dfrac{|V_C|}{|V_G|} = \dfrac{\omega L_S}{R_S} = Q$ of the inductor being measured.

The Q-meter can be satisfactorily substituted by a conventional bridge or a LCR bridge; in this case, the perfect resonance condition is not necessary and the residual admittance measured by the bridge can be employed in a balance equation (Rayner et al., 1980; Yonenaga and Nakamura, 2004).

4.9.2 Capacitance

Resonance methods are of interest in the measurement of gas-dielectric capacitors as high-frequency impedance standards. The basic assumption of the method is that the capacitance dependence $C(f)$ on f of gas-dielectric capacitors is caused by LC stray parameters of electrodes the internal connections to the defining ports, Sec. 8.3.1.

In practice, $Z(f)$ spectra measured on the standard according to the appropriate definition chosen show a number of series and parallel high-Q self-resonance modes, typically widely spaced in frequency. The lowest self-resonance frequency f_0 is typically at several tens of MHz, see Fig. 8.24; and when $f \ll f_0$, the second-order expansion in f of the capacitance $C(f)$ deviation from its low-frequency value C_0 (Eq. (8.1)) can be written as

$$C(f) \approx C_0 \left[1 \pm \left(\frac{f}{f_0} \right)^2 \right] \tag{4.22}$$

(sign $+$ is valid if there is a series resonance at frequency f_0, sign $-$ if a parallel resonance occurs).

The direct resonance measurement involves shorting out the capacitor with an external bar of known inductance. The resonant frequency can be observed by the absorption spectrum of an external electromagnetic field. In Jones (1963) and Free and Jones (1992) the measurement is performed on two-terminal capacitors with a grid-dip meter. The technique can be extended to three-terminal and four terminal-pair capacitors (Jones, 1980).

A "virtual" resonance measurement can be performed by assuming a simple series LC model of the capacitor; L is measured directly by opening the capacitor and shorting out its plates (Hanke and Dröge, 1987; Nakase, 1988; Klionsky et al., 2002) with a short conducting clip.

4.10 Mutual inductance measured as self-inductance

The natural definition of a mutual inductance M is that of a 2P, or 4T standard (Sec. 1.6.4.1).

A 2T meter can be employed for the measurement of M by arranging the windings in series, Fig. 4.34(a); an inductance measurement L_S is obtained. The connections are rearranged to obtain an anti-series, Fig. 4.34(b) and a

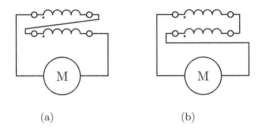

(a)   (b)

Figure 4.34
(a) Series connection of the two windings of a mutual inductor, measured as self-inductance L_S by meter M. (b) Antiseries connection of the same windings; meter M measures L_A.

second measurement L_A is taken. Then,

$$M = \frac{1}{4}\left(L_S - L_A\right). \tag{4.23}$$

The accuracy of the method is limited by the inductance of the connection, and by stray capacitances between windings (which are at different potentials in the two configurations).

A corresponding parallel/antiparallel measurement can be employed; however, the expression corresponding to Eq. (4.23) is more complicated.

4.11   Network analysis

In the past, impedance measurements at radio and microwave frequencies have been conducted with a variety of dedicated methods and circuits (see Somlo and Hunter, 1985, for a review). Nowadays, the instrument of election for these measurements is the *network analyzer*. Both *scalar* (SNA) and *vector* (VNA) network analyzers exist, an obvious reference to the capability of measuring only magnitude of electromagnetic quantities, or both magnitude and relative phase.

Network analyzers include several components; the following are always present:

signal source supplies the sinewave stimulus for the measurement. Usually the instrument can be programmed to perform *frequency sweeps* covering a bandwidth (either defined with "start" and "stop" frequencies, or "center" and "span" frequencies). Often *power sweeps* at a fixed frequency can be also programmed.

detectors, which measure the signal at their input ports. Scalar detectors (Sec. 3.2.4), such as diode detectors, give a low-frequency output proportional to the power at their input signal. Vector detectors are complex devices, which, in addition to the signal at the input port, require an additional *reference* signal, provided by a local oscillator locked to the signal source frequency f. In most vector detectors, the local oscillator frequency f_{local} is shifted of an $IF = f - f_{local}$ called the *intermediate frequency*. The input port and the local oscillator are connected to a *mixer*, which output at frequency IF, is analyzed (usually by sampling and digital signal processing) to recover amplitude and phase information about the signal under detection.

power splitters, or *power dividers*, Fig. 4.35(a), are three-port devices; the power at the input port is split (evenly, or in a given proportion) on the two output ports. The two circuits at the output ports are in this way energized with coherent signals. The simplest power splitter is the *resistive tee*, Fig. 4.35(b), which, however, has a significant power loss.

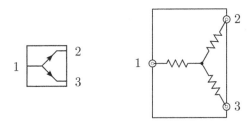

(a) Power splitter symbol. (b) Resistive tee power splitter.

Figure 4.35
Power splitter. (a) A power splitter symbol for schematic diagrams. 1 is the input port, 2 and 3 the output ports. (b) Resistive tee realization of a power splitter. Each resistor has resistance $R = \frac{1}{3}Z_0$.

directional couplers are four-port devices, where the coupling between ports is dependent on the direction of the signal power flow, and are employed to separate forward and reflected power waves in a network. Ports may be named *input, through, coupled,* and *isolated* port; the behavior is shown in Fig. 4.36.

In an ideal directional coupler, signal entering the input port is splitted in a given proportion between through and coupled ports, like in a power splitter; if a signal enters the through port, it is instead coupled to input and isolated ports. Isolated port is often just terminated with a matched load, so the power routed to the isolated port is dissipated.

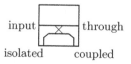

Figure 4.36
Directional coupler symbol.

A minimal network analyzer has one signal source, three detectors marked R (for *reference*), A and B, and a *test set* of splitters and directional couplers, which can be integrated in the same box or provided as a separate object. Fig. 4.37 shows the block schematics of such analyzer with a *transmission-reflection* (T/R) test set, suitable for the direct measurement of 1P, 2P, and 4C impedances.

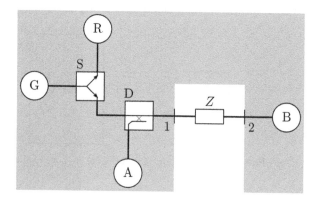

Figure 4.37
Block schematics of a simple network analyzer (gray area) with a transmission-reflection test set, connected to a 2P impedance Z. All connections are coaxial.

The splitter in the T/R test set divides the signal power; a fraction is measured by detector R, the other goes through a directional coupler D to instrument port 1, connected to the DUT (here a 2P impedance Z). The signal transmitted by the DUT to port 2 is measured by detector B. The signal reflected at port 1 is routed by the directional coupler to detector A. In a vector network analyzer, detectors R, A, B give complex-valued readings R, A, B. The estimation of the parameter matrix is given, after taking into account instrumental parameters and the results of *calibration*, Sec. 4.11.1, by ratios A/R and B/R of the readings.

4.11.1 Calibration

The *calibration* of a network analyzer (Rumiantsev and Ridler, 2008) assumes a *error model* of the analyzer, a mathematical description where the deviations of the real analyzer behavior from an ideal one are associated to a number of *error terms*. The calibration ask for a *calibration set*, consisting of a number of standards (typically one-port or two-port passive devices) see Sec. 8.8; the standards are sequentially connected to the analyzer and measured; the analyzer readings are saved in memory. The calibration algorithm employs such readings to fit the error model and quantify the error terms. After the calibration, the algorithm can correct the readings of a measurement performed on an unknown device, to improve the measurement accuracy.

The error terms belong to different classes. Those related to signal leakage through unwanted paths are called *directivity* and *crosstalk*; those related to unwanted reflections at the test ports are called *source mismatch* and *load mismatch*; errors caused by the frequency response of the receivers are called *transmission tracking* and *reflection tracking*.

Several calibration schemes exist; the most popular is called *SOLT* (*short-open-load-thru*) and employs four standards: a short termination ($Z \approx 0$), an open termination ($Y \approx 0$), a load ($Z \approx Z_0$) and a *through*, where the analyzer ports are connected directly (if provided with hermaphrodite connectors or a male/female connector couple), or with a short adapter.

Calibrated short and open standards are difficult to obtain at higher frequencies, or in noncoaxial measurement conditions (e.g., for measurements on silicon wafers): the *TRL* (*through-reflect-line*) calibration employs, in addition to a through, an uncalibrated high-reflection device (open or short) and a transmission line of calibrated electrical length (see also Sec. 9.2.2).

The expression of uncertainty of network analyzer measurements is presently matter of scientific debate (Williams et al., 2003; Stumper, 2005).

4.11.2 Six-port reflectometer

Measurement of impedance as a complex quantity ask for a vector network analyzer, which includes vector detectors. An alternative instrument that employs only (much simpler) power detectors is the *six-port reflectometer*.

The basic idea (Hoer, 1972) is similar to that of the three-voltage method (Sec. 4.3): if both the magnitudes of two phasors, and the magnitude of the sum (or difference) of the same phasors are measured, then (apart a sign ambiguity) their relative phases can be obtained.

The six-port reflectometer is constructed around a six-port passive linear network, which is fully defined by its 6×6 scattering parameter matrix. The measurement circuit, Fig. 4.38, includes the network, a generator at port 1, the impedance to be measured Z at port 2, and four power detectors P_i at ports $i = 3 \dots 6$.

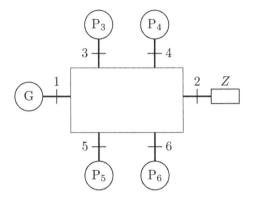

Figure 4.38
A six-port reflectometer. Adapted from Engen (1977a, Fig. 2).

The instrument reading is given by the set of four power detector readings P_i, $i = 3\dots6$.

The normalized power readings $p_i = \dfrac{P_i}{P_6}$ ($i = 3\dots5$) are related to the reflection coefficient Γ (Sec. 1.9) of Z_{X} by the three nonlinear expressions

$$p_i = \frac{P_i}{P_6} = C_i \cdot \left| \frac{\Gamma - q_i}{d\,\Gamma + 1} \right|^2 \qquad i = 3\dots5, \tag{4.24}$$

where constants C_i (real), q_i (complex), and d (complex) are dependent on the six-port network, and are determined by a calibration phase, by substituting Z at port 2 with known reference standards. System (4.24), called the *working equations* of the reflectometer, give an overdetermined nonlinear system that is solved (often graphically on the Smith chart, Sec. 1.9) to obtain Γ (hence, Z) from P_i, $i = 3\dots6$.

Although in principle any nondegenerate six-port network can be employed, its scattering matrix has a decisive influence on the measurement accuracy of the resulting six-port reflectometer in the measurement of a particular Z_{X}. Criteria for the design of an appropriate network are given by Engen (1977b).

4.11.3 Time-domain reflectometry

Time-domain reflectometry, TDR, is a particular method of one-port network analysis, which employs pulses (or steps) as test signal instead of sinewaves. The technique is widespread in the localization of faults in electrical wiring. The test signal travels from the test port to the cable and is reflected at discontinuities in the characteristic impedance (e.g., a short circuit in a cable). The separation between incident and reflected signals is given by the round-trip travel time of the pulse between the test port and the impedance discontinuity; no directional couplers are necessary.

For localization purposes, measurements can be performed directly on the time axis of an oscilloscope connected in parallel to the test port. Quantitative impedance spectroscopy require fast sampling and FFT analysis, see Sec. 5.5.1.

TDR can achieve very large measurement bandwidths, up to frequency domains (terahertz and optical frequencies), where pulse generators can be more easily available than swept-sinewave generators.

4.11.4 Network analysis of impedance standards

Impedances defined as n-terminal pair standards $Z_{n\mathrm{P}}$ (Sec. 2.2) can be considered a n-port passive network (Sec. 1.6.2). Network analysis (Sec. 4.11) permits the measurement of $\boldsymbol{Z}$, $\boldsymbol{Y}$, or $\boldsymbol{S}$ matrix elements; $Z_{n\mathrm{P}}$ can then be evaluated with the relations given in Tab. 2.1.

4.11.4.1 Indirect measurement of Z

The method, called sometimes *Z-matrix method*, was developed by Suzuki (1991) and Aoki et al. (1998), with the goal of measuring the capacitance and dissipation factor of four terminal-pair gas-dielectric capacitors. The measurement of $\boldsymbol{Z}$ is achieved by *one-port* network analyzer measurements only. The diagonal elements Z_{ii} of $\boldsymbol{Z}$ are measured directly with the analyzer, all ports $k \neq i$ being left open (as requested by the definition of $\boldsymbol{Z}$, Sec. 1.6.3.1). Off-diagonal elements Z_{ij}, $i \neq j$ are measured indirectly, with the relation

$$Z_{ij} = Z_{ji} = \pm\sqrt{Z_{jj}(Z_{ii} - Z_{ii(\mathrm{s}j)})}, \qquad (4.25)$$

where $Z_{ii(\mathrm{s}j)}$ is the impedance at port i, when port j is shorted (and remaining ports are left open).[5]

By combining Eq. (4.25) with the relation $Z_{4\mathrm{P}} = \dfrac{Z_{34}Z_{21} - Z_{24}Z_{31}}{Z_{31}}$, given in Tab. 2.1, the measurement equation

$$Y_{4\mathrm{P}} = \sqrt{\frac{Z_{11} - Z_{11(\mathrm{s}3)}}{Z_{22}}}$$
$$\cdot \left[\sqrt{(Z_{11} - Z_{11(\mathrm{s}2)}) \cdot (Z_{44} - Z_{44(\mathrm{s}3)})} - \sqrt{(Z_{11} - Z_{11(\mathrm{s}3)}) \cdot (Z_{44} - Z_{44(\mathrm{s}2)})}\right]^{-1}$$

can be written (Suzuki, 1991).

One-port measurements with the network analyzer can be not sufficiently accurate for a direct calculation of $Y_{4\mathrm{P}}(f)$ from $\boldsymbol{Z}(f)$ at each frequency point of interest. To overcome such difficulty, the measurements of $\boldsymbol{Z}(f)$ can be performed in a much larger bandwidth than the frequency range of interest, often going to hundreds of MHz. Afterword, $\boldsymbol{Z}$ is employed to identify

---

[5]Eq 4.25 assumes reciprocity of $Z_{n\mathrm{P}}$.

a lumped-parameter model of the capacitor being measured (Yonekura and Wakasugi, 1990), like the one in Fig. 4.39 (Avramov-Zamurovic et al., 2000). Koffman et al. (2000) give an evaluation of the uncertainty of the method for the measurement of Agilent 16380A capacitors. More recently, Suzuki (2009) described an approach which mixes resonant and network analysis techniques.

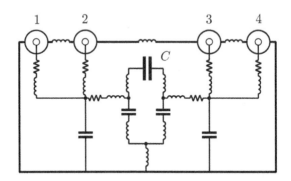

Figure 4.39
Lumped-parameter model of the Agilent Tech. 16380A standard capacitor set. The model reflects the physical construction of the standards. Adapted from Yonekura and Wakasugi (1990).

4.11.4.2   Direct measurement of Z

Awan et al. (2004) propose a direct measurement of Z with a voltamperometric technique. To measure the $Z_{ij} = V_i/I_j$ element of Z, a current generator injects a current I_j in the port j; the voltage V_i is measured with a vector voltmeter. The input and output impedances of the instruments, typically $50\,\Omega$, are raised (at the expense of sensitivity) with a high impedance in series. Z is measured apart from a frequency-dependent gain factor (but having the same value for all Z_{ij} elements for a given frequency), which requires calibration. In the measurement of a gas-dielectric capacitor, the gain calibration can be avoided if only the resonance frequency (Sec. 4.9) is of interest.

4.11.4.3   S-matrix measurements

Direct measurement of the scattering parameter matrix S is possible with two-port VNA. We consider here the more complex case of a 4P standard; measurement of 2P and 4C standards are also possible. Tab. 2.1 gives the relation between Z_{4P} and S (Callegaro and Durbiano, 2003), here reprinted for convenience:

$$
\begin{aligned}
Z_{4P} = {} & 2Z_0[S_{21}S_{34} - S_{31}S_{24}] \\
& \cdot [S_{31} + (S_{21}S_{32} - S_{31}S_{44} - S_{31}S_{22} + S_{41}S_{34} - \\
& - S_{21}S_{32}S_{44} + S_{21}S_{34}S_{42} + S_{31}S_{22}S_{44} - \\
& - S_{31}S_{42}S_{24} - S_{41}S_{34}S_{22} + S_{41}S_{24}S_{32})]^{-1}.
\end{aligned}
\tag{4.26}
$$

The measurement can be performed in a single run with a four-port network analyzer. With a two-port analyzer, S elements have to be measured in sequence by connecting to the analyzer two standard terminal-pairs at a time, and by terminating the remaining terminal-pairs to matched loads, see Fig. 4.40.

The technique has been sometimes called *S-matrix method*. Fig. 4.41 and 4.42 show an example of measurement, see Callegaro and Durbiano (2003) for details. An evaluation of the uncertainty of the method with Monte Carlo propagation has been proposed (Callegaro, 2006).

Özkan et al. (2007) conjugated S measurements of Z_{4TP}, performed on gas-dielectric capacitors (Agilent mod. 16380A series) with an identification of the corresponding lumped-parameter model, Fig. 4.39.

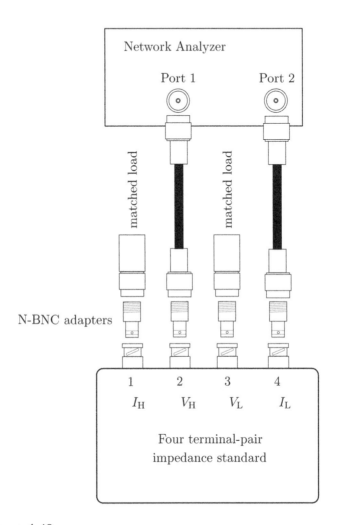

Figure 4.40
Example of connection of a 4P standard terminated with BNC male connectors
to a two-port network analyzer with N connections. The particular connection
shown permits the measurement of S_{24}, S_{42}, S_{22}, and S_{44}.

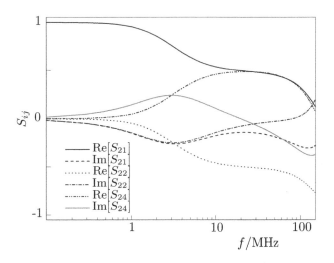

Figure 4.41
Results of S measurement on a 1 nF capacitor (Agilent mod. 16384A),
with the setup of Fig. 4.40. A set of nine spectra was obtained; the subset
S_{22}, S_{23}, S_{24} is shown, separating real and imaginary parts. Other spectra of
the set, not shown, have similar appearances.

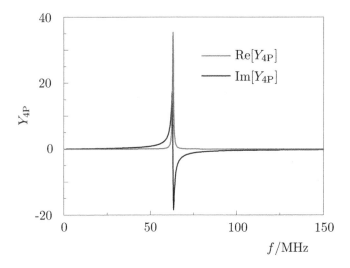

Figure 4.42
The 4P admittance Y_{4P}, as computed by means of Eq. (4.26) from the S measurements of Fig. 4.41. The frequency, $\approx 65\,\text{MHz}$, of the sharp LC resonance
can be employed the input datum of Eq. (4.22).

5

Going digital

CONTENTS

As in many other test and measurement fields, the use of *mixed-signal electronics* in impedance measurement setups is increasing. Mixed-signal electronics give the interface between the analog part of the instrument (analog electronics and electromagnetic components), which drives the excitation signal and senses voltages and current on the impedance under measurement, and the digital signal processing unit.

5.1   Sampling

An analog signal $x(t)$ is naturally continuous in time t. *Sampling* is the reduction of $x(t)$ to a corresponding sequence of *samples*, time-discrete values

$x[k] = x(k\,T_{\mathrm{S}})$, where k is integer and T_{S} is the *sampling interval*. The inverse of sampling time is the *sampling rate* or *sampling frequency*, $f_{\mathrm{S}} = T_{\mathrm{S}}^{-1}$.

5.1.1   The sampling theorem

The Nyquist-Shannon *sampling theorem* states that if:

- $x(t)$ is *band-limited* to bandwidth B (that is, if its power spectral density is zero above frequency B),

- $f_{\mathrm{S}} \geq 2B$,

then $x(t)$ is completely determined by its complete sample set $\{x[k]\}$, and the *Whittaker-Shannon interpolation formula* applies:

$$x(t) = \sum_{k=-\infty}^{\infty} x[k]\,\mathrm{sinc}\left(\frac{t - kT_{\mathrm{S}}}{T_{\mathrm{S}}}\right). \tag{5.1}$$

5.1.2   Quantization

Sampling is associated with *digitization*, the representation of samples in a numerical form.

The numerical representation of a sample is obtained by *quantization*, which reduces the infinite possible values of $x[k]$ to a finite set of coding words. The reduction of available information due to quantization results in the generation of *quantization noise*. As first approximation, an ideal n-bit converter has a quantization signal-to-noise ratio SNR $= 6.02n + 1.76$dB measured over the Nyquist bandwidth (Bennett, 1948; Kester, 2009).

5.1.3   ADCs and DACs

Mixed-signal electronics include *analog-to-digital converters* (ADC), which perform digitization of electrical signals, and *digital-to-analog converters*, which transform back digital codes in analog signals.

5.1.3.1   Signal conditioning

The typical input (output) signal of an ADC (a DAC) is a voltage; typical operating range of ADC (DAC) is between ± 1 and pm10 V. Therefore, the measurement (generation) of an electrical quantity requires a form of *signal conditioning*. Signal conditioning can include the analog conversion from (to) other electrical quantities (for example, if the original analog signal is an electrical current, signal conditioning can be performed by a transresistance amplifier), range matching, filtering, and electrical isolation.

ADC signal conditioning typically include an *anti-alias* lowpass filter, to constrain the analog bandwidth of the signal and match the Nyquist-Shannon criterion, and a *sample-and-hold* circuit, which hold the analog value for the time required by the analog to digital conversion process.

DAC signal conditioning include a *reconstruction filter*, which implements an approximation of the Whittaker-Shannon interpolation formula (5.1).

5.1.3.2 Properties of ADCs and DACs

Important properties of ADCs and DACs are the

- the maximum sampling rate;
- the signal range and available signal polarities;
- the resolution, expressed in bit (the minimum length of the digital word expressing the input/output code).

ADCs and DACs most significant specifications are:

- offset and gain errors;
- static linearity error, typically expressed as *integral nonlinearity* (maximum deviation over the range from linear behavior) or *differential nonlinearity* (maximum deviation from linear behavior between adjacent codes);
- dynamic error, and noise. On this matter, the specifications of ADCs and DACs can be given in terms of various acronyms such as SINAD (signal-to-noise-and-distortion-ratio), ENOB (equivalent-number-of-bits), SFDR (spurious-free-dynamic-range), NPR (noise-power-ratio), ACLR (adjacent-channel-leakage-ratio), and so on. Such quantities have a complex (and not always unique) definition, which can be found in specialized literature (Kester, 2005, Fig. 2.48).

5.2 Direct digital synthesis

Direct digital synthesis (DDS) is a technique to employ digital data-processing blocks and DAC(s), as a means to generate one (or several) frequency- and phase-tunable periodic output signal(s).

5.2.1 DDS principle

In its simplest form, a direct digital synthesizer can be implemented as shown in Fig. 5.1. Digital events occur at rate f_{CK} of a precision reference clock signal. At each clock tick, a m-bit tuning word M is added to a m-bit register, called *phase accumulator*. The phase accumulator content is shifted by a user-programmable phase shift, and sent to a phase-to-amplitude converter,

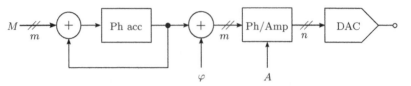

Figure 5.1
Block schematics of a direct-digital-synthesis (DDS) generator. The DDS is controlled by frequency word M, phase word φ, and amplitude word A; Ph acc is m-bit phase accumulator; Ph/amp is the m-to-n bit phase-to-amplitude converter.

which converts the m-bit input phase in a n-bit output amplitude ($n \leq m$); the conversion, which might accept also an amplitude control signal, can be performed by a real-time function calculation algorithm, or with the help of a lookup table stored in memory. The amplitude signal is sent to a DAC (possibly through a memory buffer), which output is available for signal conditioning.

Before entering the signal conditioning stage (which usually includes filtering) the DDS output is a staircase approximation of a periodic waveform, having frequency f_{OUT}

$$f_{\text{OUT}} = \frac{M}{2^m} f_{\text{CK}}.$$

Each approximation of a period is a staircase waveform, given by $\dfrac{2^m}{M}$ steps[1] chosen over 2^n possible amplitude levels.

If the application ask for more than one output, the phase accumulator can be processed in parallel by multiple channels, see Fig. 5.2. In the resulting polyphase generator, outputs are locked in frequency and with a known phase relationship.

The advantages of DDS over analog synthesis are

- full digital control of the output analog waveforms, permitting high-speed amplitude, frequency and phase modulation schemes and real-time modifications of the output waveform;

- high f_{OUT} accuracy, which is strictly related to the accuracy of f_{CK};

- high f_{OUT} resolution (often down to µHz level, depending on the phase word length m);

---

[1] In general $\dfrac{2^m}{M}$ is not an integer; therefore, the approximations are different from one period to the next. f_{OUT} is the repetition frequency of the original waveform; the repetition frequency of the particular approximation sequence can be much smaller.

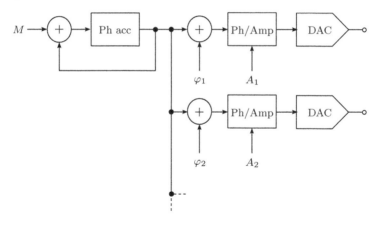

Figure 5.2
The extension of the DDS principle of Fig. 5.1 to a polyphase generator, controlled by amplitude signals $A_1, A_2, \ldots$ and phase signals $\varphi_1, \varphi_2, \ldots$.

- for multi-output DDS synthesizers, a high resolution of amplitude and phase relationships between the channels.

A typical application of DDS is the generation of sinewaves. The output waveform of a DDS is affected by a number of spurs, which have different origin (Torosyan and Wilson, 2005):

- sampling. The spectrum of a sampled output has *images* at frequencies $f_{im} = f_{CK} \pm f_{OUT}, 2f_{CK} \pm f_{OUT}, \ldots$, which amplitudes relative to fundamental are given by the $\text{sinc}(\pi f_{im}/f_{CK})$ weight function.

- quantization. An n-bit quantization generate harmonics of f_{OUT} frequency, and a quantization noise floor (Sec. 5.1.2).

- phase truncation. If the phase to amplitude conversion is performed with precalculated waveform values stored in a memory, the m-bit phase can be truncated to p bits in order to reduce the lookup table to 2^p bits. This creates additional spurs, which amplitude is strongly dependent on the particular frequency chosen.

Such spurs are intrinsic to the DDS generation principle. A number of additional spurs and output noise can occur because of DAC static and dynamic errors, Sec. 5.1.3.2.

5.2.2 DDS implementations

Various kinds of DDS implementations exist, from large-scale integration on single chips, typically employed in radio-frequency circuits (f_{OUT} can go up to GHz range) to software implementations on personal computers provided with DAC output boards.

When employing DDS generators in electrical instrumentation, care must be taken to keep under control the harmonics and spur level. The output DAC performance can be degraded by analog signal processing components (buffer amplifiers, isolation transformers, etc.) and the properties of the load to be driven (for example, highly reactive loads can resonate at spur frequencies).

Spurs can cause systematic errors when wideband instruments (e.g., RMS voltmeters) are employed in the measurement. Spurs limit also synchronous detector sensitivity.

In polyphase DDS synthesis, amplitude and phase relationship between different outputs can be arbitrarily programmed; the programming accuracy is related to the properties of the DACs and the analog processing electronics. Typically, the amplitude *ratio* between two outputs is more accurate than the absolute amplitude of each output, in particular if the same reference voltage is employed for both DACs. The phase *difference* error is related to different phase rotations in the analog channels, and is smaller than the absolute phase rotation of each channel.

Amplitude and phase ratio can be improved by off-line adjustment with so-called *autozero* circuitry (Turgel and Oldham, 1978; Hess and Clarke, 1987). High accuracy can be obtained with an active control loop based on an auxiliary output, which periodically shuttles between that of main outputs (Cabiati and Pogliano, 1987).

5.2.3 Josephson DDS

A new class of DDS synthesizers of particular interest for impedance metrology is based on Josephson DACs (see Jeanneret and Benz, 2009, for a review).

A Josephson junction is a device consisting of two superconducting electrodes separated by a thin insulating barrier. The junction has a nonlinear behavior: if set at an appropriate dc current bias and irradiated with microwave power at frequency f_d, it develops an average voltage

$$V = \frac{h}{2e} n f_d = \frac{n f_d}{K_J},$$

where $K_J = \dfrac{2e}{h}$ is the *Josephson constant*, see App. C, and n is a small integer called the step number. A typical application can employ $f_d \approx 75\,\text{GHz}$, and $n = \pm 1$, giving $V \approx 150\,\mu\text{V}$. Modern microfabrication techniques permit to obtain Josephson junction arrays composed of up to 500 000 (Yamada et al., 2010) individual elements in series, to achieve output voltages to 10 V or larger.

In so-called *overdamped* junctions, n can be switched (typically between values $-1, 0, 1$) by choosing the corresponding dc bias current I; the resulting voltage output is switched in magnitude and polarity. Fig. 5.3 shows an *I-V* characteristic of an overdamped junction under microwave irradiation.

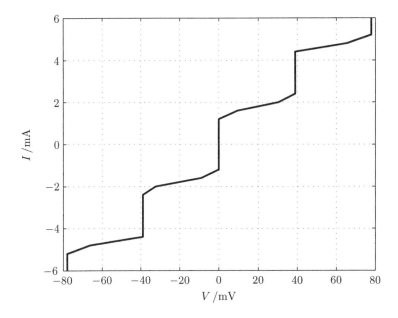

Figure 5.3
The *I-V* characteristic of an overdamped Josephson junction under microwave
irradiation. $n = 0$ and ± 1 voltage steps are visible, and $n = \pm 2$ steps can be
partially seen. Actual characteristic is measured at $f \approx 73\,\text{GHz}$ on an array of
256 equal junctions in series, so voltage axis is expanded 256 times respect to
a single junction. Data courtesy of N. De Leo and M. Fretto, INRIM.

If a Josephson array is electrically divided into binary segments, each one
composed of $1, 2, 4, \ldots, 2^n$ junctions, and each segment is individually current
biased, a binary-weighted DAC is obtained, see Fig. 5.4.

The Josephson DAC has no static errors. Its dynamical errors, however,
are influenced

- by the dynamical properties of the current bias source, presently realized
  with conventional electronics, since when switching between current bias
  values corresponding to different n values, a non-quantized transient output voltage occurs;

- by the electrical properties of the bias and output wiring, necessarily long
  ($>1\,\text{m}$) because of the need of a cryogenic environment;

- by the load impedance.

A Josephson DAC can be employed for DDS synthesis. Analog output
electronics (reconstruction filter, buffer amplifier) is typically not present, to

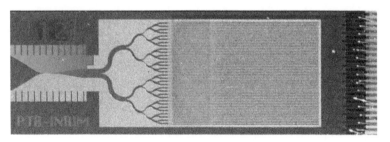

Figure 5.4
A Josephson junction binary array chip, composed of 8192 superconducting-normal metal-insulator-superconductor (SNIS) junctions, suitable for operation at temperatures up to 7 K (Lacquaniti et al., 2011). The junctions are geometrically arranged over 32 parallel strips of 256 junctions each, and driven by a microwave at 70 GHz coupled by the antenna and tree splitter on the left. Electrically, the junctions are connected in series and arranged in binary sections, which can be individually current polarized (by the electrical contacts on the right) to select positive, negative, or zero voltage steps, resulting in a 13 bit+sign DAC with $\approx \pm 1.2$ V fullscale.[3]

maintain the accuracy of quantized output steps; therefore, harmonics and images are present in the output spectrum.

A two-phase Josephson DAC has been recently employed in the realization of 2P and 4P resistance ratio bridges working in the audio-frequency range, with uncertainties of parts in 10^8 (Lee et al., 2010, 2011).

5.3   Digital signal analysis

Sampling and digital processing of signals in a measurement setup is called *digital signal analysis*, here briefly recalled.

5.3.1   Fourier analysis

The frequency domain representation of a generic signal $x(t)$ is given by its amplitude spectrum (Sec. 1.7.1)

$$X(f) = \int_{-\infty}^{\infty} x(t) \exp\left(-2\pi \mathrm{i} f t\right) \mathrm{d}t$$

The frequency analysis of sampled signals $x[k] = x(kT_\mathrm{S})$, where $k = 0 \ldots N - 1$ and N is the sample length, is performed with the so-called *discrete Fourier transform* DFT, a time-discrete analogue of the amplitude spectrum:

$$X[m] = \sum_{k=0}^{N-1} x[k] \exp\left(-\frac{2\pi i}{N} mk\right),$$

where $m = 0 \ldots N - 1$ are associated to *frequency bins* of width f_S/N over the frequency set $\{mf_S/N\}$, called *bin centers*.

The efficient calculation of the DFT is performed with *fast Fourier transform* (FFT) algorithms.

In general, even if Nyquist-Shannon condition on sampling rate is satisfied, the relation between the DFT of the sampled signal and the Fourier transform of the original signal is distorted by an effect called *spectral leakage*: it appears as if some energy has "leaked" out of the original signal spectrum into other frequencies. Spectral leakage can be reduced, but not eliminated, by applying appropriate *window functions* to the original samples (Percival and Walden, 1993, Sec. 6.4.).

5.3.2 Synchronous sampling

Synchronous or *coherent* sampling of a periodic signal $x(t)$ (having period T and frequency $f = 1/T$) refer to the collection of N samples, with a sampling time T_S such that an integer number of periods $P < N/2$ of the signal perfectly fits in the sampling time window: $PT = NT_S$.

Synchronous sampling is of interest because it eliminates spectral leakage: every harmonic Hf of the signal frequency f (where $H = 1, 2, \ldots$ is the harmonic order) up to sampling frequency corresponds to a bin center $m_H = HP$ of the DFT $X[m]$. If $x(t)$ is a sinewave, amplitude and phase information is given by the complex number $X[P]$.

The measurement of impedance at frequency f can be reduced to the determination of the complex ratio $R = X(f)/Y(f)$ of two isofrequencial signals $x(t)$ and $y(t)$. For the case of synchronous sampling, the ratio can be computed from their samples $x[k]$, $y[k]$ as $R = X[P]/Y[P]$.

5.3.3 Asynchronous sampling

When synchrounous sampling conditions are not met, spectral analysis with DFT is complex and prone to systematic errors. In literature, asynchronously sampled signals are analyzed by *fitting*: it is assumed that the signal belong to a particular class of waveforms, called *model*; a nonlinear regression analysis on the available signal samples return the waveform parameters of interest.

We limit the discussion to *sine fitting*, which assumes the sinusoidal model[4]

$$x_m(t) = A\cos(2\pi ft) + B\sin(2\pi ft) + C, \tag{5.2}$$

---

[4]Other forms of sinusoidal model, e.g., $x_m(t) = A\cos(2\pi ft + \phi) + C$ of course can be considered.

where A, B, C, f are the parameters of the model. The samples $x[k] = x(kT_S)$ and model (5.2) are the input of the sine-fitting algorithm, which outputs estimates $\hat{A}$, $\hat{B}$, $\hat{C}$, and $\hat{f}$ of the original sampled waveform.

Sine-fitting algorithms based on least-squares analysis have been considered in standards (IEEE 1057, par 4.6; IEEE 1241, par 4.1) and are the subject of active research (Händel, 2000; Bilau et al., 2003; Fonseca da Silva et al., 2004; Kollar and Blair, 2005; Chen, 2007; Chen and Xue, 2008).

As for synchronous sampling, impedance measurement requires considering two sample sets $x[k]$ and $y[k]$ of isofrequencial signals $x_m(t)$ and $y_m(t)$;

$$x_m(t) = A_x \cos(2\pi ft) + B_x \sin(2\pi ft) + C_x,$$
$$y_m(t) = A_y \cos(2\pi ft) + B_y \sin(2\pi ft) + C_y; \tag{5.3}$$

the sine-fit analysis can be performed on both sets (Ramos et al., 2004) $x[k]$ and $y[k]$, and the complex ratio R estimated by

$$\hat{R} = \frac{\hat{A}_x + i\hat{B}_x}{\hat{A}_y + i\hat{B}_y}.$$

5.4 Digital impedance bridges

The outputs of a polyphase DDS generator can be employed in bridge schemes.

A simple voltage ratio bridge (Sec. 4.4) can be implemented with a two-phase generator, see Fig. 5.5(a). The two-terminal bridge can compare impedances having arbitrary magnitude and phase relationships. Implementations of the principle of Fig. 5.5(a) (Bachmair and Vollmert, 1980; Helbach et al., 1983; Waltrip and Oldham, 1995) define impedances as 2P standards.

More complex bridges may require generators having more than two phases, also to achieve auxiliary equilibria. For example, Fig. 5.5(b) shows a 4T voltage ratio bridge (Ramm, 1985), and Fig. 5.6 a 4P one.

5.4.1 Digitally assisted impedance bridges

The accuracy of digital bridges is directly dependent on the accuracy of the generators employed, since the equilibrium setting determines the measurement. It is interesting to combine the advantages of digital bridges with the accuracy of classical transformer bridges. We can speak of *digitally assisted bridges* for this class of instruments.

An example of a digitally assisted ratio bridge is shown in Fig. 5.7, for the comparison of like impedances Z_1 and Z_2. The bridge is energized with voltage E_a by phase 1; the inductive voltage divider has a fixed and known ratio k, which can be altered by the small voltage E_b derived from generator

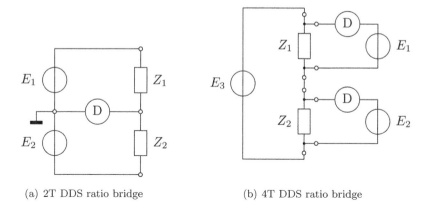

(a) 2T DDS ratio bridge   (b) 4T DDS ratio bridge

Figure 5.5
Voltage ratio bridges implemented with DDS generators. (a) 2T voltage ratio bridge implemented with a two-phase DDS generator. (b) 4T voltage ratio bridge implemented with a three-phase generator. Adapted from Helbach and Schollmeyer (1987).

phase 2 (for example, with an injection feedthrough transformer) to achieve equilibrium on detector D. If $\epsilon = \dfrac{E_b}{E_a} \ll 1$, the equilibrium condition is

$$\frac{Z_1}{Z_2} = \frac{1 - k + \epsilon}{k - \epsilon} \approx \frac{1 - k}{k} + \frac{1}{k^2}\epsilon;$$

the bridge accuracy is determined by the IVD, since the generator accuracy affects only the correction term ϵ.

Digitally assisted bridges are of particular interest when one considers four terminal-pair impedance comparisons, since a large part of adjustments needed in classical bridges do not enter the measurement equation, but are necessary for impedance definition. Fig. 5.8 shows an example of a 4P current comparator impedance bridge (Sec. 4.5.1).

Digitally assisted bridges reaching accuracies suitable for primary metrology have been published (Muciek, 1997; Corney, 2003; Trinchera et al., 2009; Callegaro et al., 2010a).

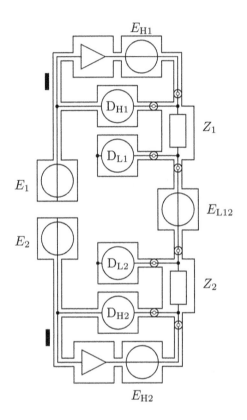

Figure 5.6
A 4P DDS ratio bridge. The bridge compares impedances Z_1 and Z_2. E_1 and E_2 are the reference voltages, for which the relation $\dfrac{E_1}{E_2} = \dfrac{Z_1}{Z_2}$ apply. Buffer amplifiers energize the high-current port of Z_1 and Z_2; correction voltages E_{H1}, E_{H2} are controlled by detectors D_{H1}, D_{H2}. D_{L1} and D_{L2} control reference ratio E_1/E_2 and the Kelvin arm E_{L12}. Adapted from Pogliano (1991, Fig. 4.3).

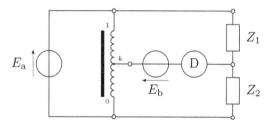

Figure 5.7
Digitally assisted ratio bridge.

5.5 Sampling impedance meters

Sampling methods can be applied to several impedance measurement methods. A simple configuration is the impedance meter of Fig. 5.9, implementing *I-V* method (Sec. 4.1) for the measurement of a 2P impedance Z.

Fig. 5.10 shows a 4T sampling vector ratiometer. Voltage drops on Z_1 and Z_2 are coupled, with differential front-end amplifiers, to a two-channel sampler (ADC1 and ADC2). In case of synchronous sampling, the excitation generator of the bridge can be a DDS generator, with a clock derived from the same source of the sampler.

If a longer measurement time is acceptable, and the measurement conditions are sufficiently stable, the two-channel sampler input connections to Z_1 and Z_2 can be periodically switched. The average of the two readings taken in the two switched conditions compensates for amplitude and phase offsets between the two channels. A single-channel sampler periodically switched between Z_1 and Z_2 also gives such compensation, but ask for a precise synchronization between the two sample sets (Ramm and Moser, 2001, 2005; Overney et al., 2008; Overney and Jeanneret, 2010).

Asynchronous sampling setups asking for high-power excitations (e.g., for measuring very low impedances) can be driven by ac power line with a decoupling transformer (Callegaro et al., 2010b).

An example of an on-chip sampling impedance meter is shown in Fig. 5.11.

5.5.1 Impedance spectroscopy

The measurement of $Z(f)$ in a wide frequency range is called *impedance spectroscopy* and is of interest for modeling of electrical and electronic components,

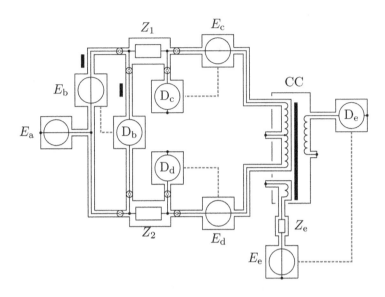

Figure 5.8
Block schematics of a digitally assisted current comparator bridge, suitable
for the comparison of 4P impedances Z_1 and Z_2. A 5-phase DDS generator
(having outputs $E_a \ldots E_e$) is employed. E_a energizes the impedances. E_b,
controlled to null detector D_b, equals voltage at PH ports of Z_1 and Z_2. E_c
and E_d, controlled to null detectors D_c and D_d, set 4P definition condition at
LP port of Z_1 and Z_2. The main equilibrium condition is sensed by detector
D_e at the detection winding of the current comparator CC, and is achieved
by the current injection given by E_e and impedance Z_e.

measurement on materials, electrochemistry (where it is called EIS, electrochemical impedance spectroscopy), and biology (Ch. 6). Impedance measurement methods described in Ch. 4 can be employed for impedance spectroscopy
if the corresponding implementation is fully automated: after the frequency
set of interest $\{f_k\}$, $k = 1 \ldots n$ is chosen, the energizing generator is set to
each f_k in sequence, the meter is adjusted, and a reading is taken. LCR meters and network analyzers include software to perform both calibration and
measurement frequency sweeps, can save readings in memory, and give $Z(f)$
representations (Sec. 1.9) on display.

The principal limitation of frequency-swept impedance spectroscopy is the
total measurement time since each frequency point measurement last several
periods. This can be unacceptable when the measurement frequency is very
low (in the Hz range or below), the frequency resolution (number of frequency
points) has to be very high, or when the total measurement time for the entire

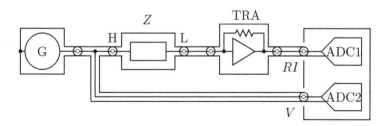

Figure 5.9
A 2P *I-V* sampling impedance meter. G energizes impedance Z and applied voltage V is sampled by ADC2. Current I is converted by a transresistance amplifier TRA to a voltage RI, sampled by ADC1.

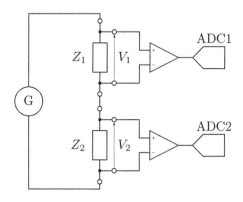

Figure 5.10
A 4T sampling vector ratiometer. G energizes impedances Z_1 and Z_2 in series; voltages V_1 and V_2 are sampled by ADC1 and ADC2.

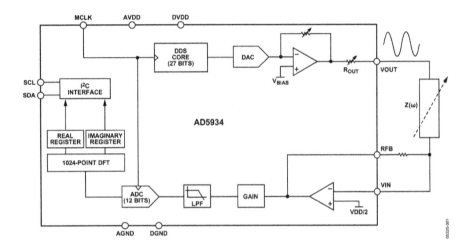

Figure 5.11
An on-chip sampling impedance meter, Analog Devices AD5934. The chip includes a sinewave DDS voltage source (composed of the DDS core, a DAC, and a buffer output amplifier with output resistance R_{OUT}) that energizes the impedance under measurement $Z(\omega)$. Current output from Z is sampled (by the chain composed of a transimpedance current to voltage converter with external resistance gain RFB, a programmable gain amplifier, an anti-aliasing filter LPF, a 12-bit ADC) and a digital signal processing system based on FFT analysis. From AD5934 datasheet, reprint permission from Analog Devices, Inc.

sweep must be very small because the impedance is unstable (typical of electrochemistry, where a chemical reaction is ongoing during the measurement).

Recalling the results given in Sec. 1.7.1, if the hypotheses given (linearity, causality, time-invariance during the measurement time) are valid, then impedance $Z(f)$ can be defined from general voltage $v(t)$ and current $i(t)$ waveforms.

An impedance measurement method can therefore employ, instead of a sinewave excitation, an excitation signal $x(t)$ (either a voltage or a current) in which amplitude spectrum $|X(f)| > 0 \quad \forall f \in \{f_k\}$, spectral analysis can be performed on resulting $v(t)$ and $i(t)$ across Z to measure $Z(f) \quad \forall f \in \{f_k\}$.

Useful test signals for impedance spectroscopy are (Herlufsen, 1984, Sec. 5):

sum of sinewaves $x(t) = \sum_{\{f_k\}} A_k \cos(2\pi f_k + \varphi_k)$. If equal amplitudes $A_k = A$, and random φ_k are chosen, for a sufficiently large $\{f_k\}$ set the resulting signal distribution approximates a Gaussian (White and Benz, 2008, Tab. 1).

white noise random or pseudorandom frequency-limited white noise has finite-amplitude spectrum for all frequencies below cutoff.

pulse, either individual pulses or periodic pulsed waveforms. A train of pulses $p(t) = \sum_{n=-\infty}^{\infty} a(t - nT)$ with repetition rate T and individual shape $a(t)$ has the spectrum $P(f) = \frac{1}{T} \sum_{k=-\infty}^{\infty} A(f - k\frac{1}{T})$, i.e., given by replicas of the spectrum $A(f)$ of $a(t)$ separated in frequency by $1/T$.

step a single step between two signal levels is a test signal easy to generate. Time-domain reflectometry (TDR, Sec. 4.11.3) employs pulse or step signals.

chirp $x(t) = \cos\left(2\pi \int_0^T f(t)\mathrm{d}t\right)$, a "sinewave" function in which instantaneous frequency $f(t)$ continuously varies with time. For an ascending linear chirp between two frequency extremes f_{start} and f_{stop}, $f(t) = (f_{\text{stop}} - f_{\text{start}})\frac{t}{T} + f_{\text{start}}$.

In practice, impedance spectrometers based on spectral analysis *sample* signals $v(t)$ and $i(t)$ and perform discrete Fourier transform analysis (Sec. 5.3.1) and are called FFT analyzers.

The scheme of Fig. 5.9 can be employed as a simple FFT impedance analyzer. An example of excitation voltage signal $v(t)$, a sum of sinewaves, is in Fig. 5.12; an example of measurement result is in Fig. 5.13.

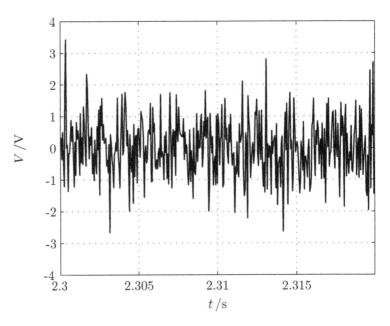

Figure 5.12
A detail of the time evolution of signal $v(t)$ employed in the impedance spectroscopy measurement. Despite the seemingly random behavior, the signal is perfectly deterministic and periodic. It is constructed from 2^{16} sinewaves, spaced in frequency of $\Delta f = \dfrac{10\,\text{kHz}}{2^{16}} \approx 0.152\,\text{Hz}$, of equal amplitudes and arbitrary phases, to give a RMS value of 1 V.

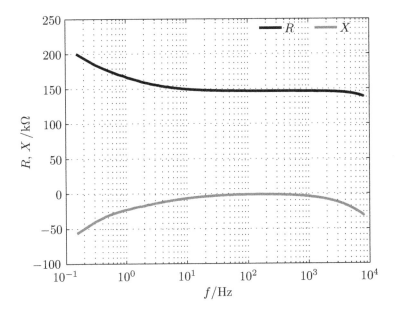

Figure 5.13
R, X plot of impedance measurement on a liquid sample (isopropyl alcohol in a coaxial cell), with the test signal of Fig. 5.12. A measurement time of $(\Delta f)^{-1} \approx 6.55\,\mathrm{s}$ is sufficient to measure a $Z(f)$ spectrum of 2^{16} frequency points.

6

Some applications of impedance measurement

CONTENTS

6.1 Measurement of electromagnetic properties of materials

Surface- or volume-specific electromagnetic properties of materials, also called density quantities (Sec. 1.10), are often inferred from impedance measurements. The property of interest is given by a measurement model in which input quantities are the impedance value(s), and the physical dimensions of the chosen geometrical configuration. The model make assumptions (e.g., material linear response, homogeneity, isotropy) whose validity has to be carefully considered.

In practice, solid materials are shaped in the form of a well-defined *sample*, which geometrical size partially defines the geometry (for example, a circular long wire having uniform and known cross section); then a *fixture* completes the definition of geometry (in the previous example, the fixture can define the measurement length) and gives the electrical connections to the impedance meter. For liquid or gaseous material, the geometry is completely determined by the fixture, which also acts as a container.

Electromagnetic properties of materials can be dependent on:

- environmental conditions (temperature, humidity, pressure);

- the properties (amplitude, frequency) of the electrical stimulus required to perform the impedance measurement. As shown in Ch. 1, impedance is a quantity defined within the linear response regime. The stimulus amplitude must be limited to prevent nonlinear response from the material.

- the *bias*, a steady-state excitation (e.g., a dc magnetic field), which sets an appropriate operating measuring condition;

- the past history, i.e., the material may show *hysteresis*.

6.1.1 Conductivity and permittivity of solid materials

Several methods are available for the measurement of conductivity/resistivity of solid materials. The choice depends on the resistivity range investigated, and the available sample shape (bar, thick plate, thin sheet, thin film). Some methods permit to measure also the electrical permittivity.

6.1.1.1 Bar method

The sample is shaped in the form of long and thin cylinder of uniform cross-section S (not necessarily a circular cylinder: a rectangular bar is a typical shape). The 4T measurement of dc or low-frequency resistance R is performed with a fixture, see Fig. 6.1, having two current contacts connected at bar extremes, and two voltage contacts in the form of knives at distance ℓ.

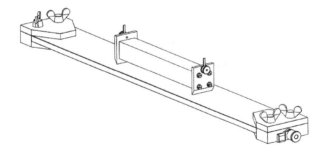

Figure 6.1
A fixture for measurement of conductivity of metallic samples, in the form of
a rectangular bar. Clamps at the end of the bar give the current contacts, and
an element with two knife edges provides the voltage contacts. ©2003 IEEE.
Reprinted, with permission, from Rietveld et al. (2003, Fig. 1).

If skin effect (Sec. 1.10.1) can be neglected, the material conductivity is
given by

$$\sigma = \frac{1}{R} \frac{\ell}{S}.$$

6.1.1.2 Eddy current method

In the most common application of the method, the sample is a slab of material
of uniform thickness. The fixture is a nonmagnetic-core cylindrical inductor of
square cross section, having a diameter smaller than the material sheet size.
The fixture is positioned with the inductor axis perpendicular to the material
sheet at a known distance, see Fig. 6.2.

When energized, eddy currents (Sec. 1.10.1) in the material, dependent
on its conductivity, alter the flux amplitude and distribution, and hence the
measured impedance $Z(\omega)$.

The measurement can be conducted at a fixed frequency (a typical value
is 60 kHz) or over a frequency range (100 Hz–100 kHz). A mathematical model
(Dodd and Deeds, 1968; Bowler and Huang, 2005) is then employed to derive
σ from $Z(\omega)$ value(s). The method is also the subject of standards (ASTM E
1004). The measurement uncertainty is usually not better than a few percent.

An alternative geometry is given by the cylindrical mutual inductor, where
the sample constitutes the inductor core (Gugan, 1997). Because of skin effect,
the effective cross section of the magnetic flux is smaller than the geometric
cross section of the core material; therefore, conductivity can be evaluated
from a mutual inductance measurement.

Eddy current measurement of conductivity assumes the homogeneity of
the measured sample. Eddy current *testing* is based on the opposite assump-

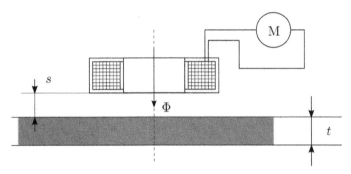

Figure 6.2
Conductivity measurement using the eddy current method. The cylindrical inductor is placed with its axis perpendicular to the material slab (having thickness t) under measurement, at a distance s, so the the magnetic flux Φ interacts with the material.

tion: the meter can detect inhomogeneities (e.g., cracks) in the material being tested. The method is employed for highly conductive metals, e.g., aluminum alloys for the aircraft industry.

6.1.1.3   Sheet method

The sample is a sheet of material of uniform thickness, placed in a fixture having cylindrical geometry as shown in Fig. 6.3(a). The guard electrode ensures both a uniform electric field in the sample volume being measured, and the collection of surface currents. The contact configuration gives a 3T impedance: the method is therefore suitable for materials having high (e.g., intrinsic semiconductors) to extremely high (insulators) resistivity. At measurement frequency ω, the conductivity σ and permittivity ϵ can be expressed in terms of the measured admittance $Y = G + jB$ as

$$\sigma = K\,G;$$
$$\epsilon = K\frac{B}{\omega}, \tag{6.1}$$

where K (m^{-1}) is the *cell constant*, which can be calibrated with a reference material or calculated as $K = d\,S^{-1}$ where d is the material thickness and S the surface area of the low-current electrode.

Connections to the fixture can be reconfigured as in Fig. 6.3(b) to measure, instead of volume properties, the surface resistivity of the material.[1]

On insulators, the applied voltage source is typically a very low-frequency

---

[1] For electrodes having circular symmetry, where the gap between them is a circular annulus with inner radius R_1 and outer radius R_2, the surface resistivity ρ_S is linked to the

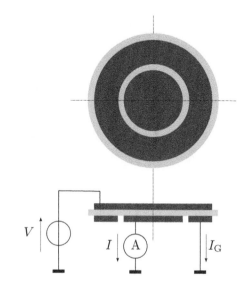

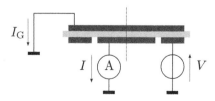

(a) Volume sheet resistivity measurement.

(b) Surface sheet resistivity measurement.

Figure 6.3
Sheet method for resistivity measurements. The fixture has a cylindrical symmetry and is shown in cross section. Voltage V is applied and current I measured; current I_G flowing in the guard electrode is not measured. (a) Measurement of volume resistivity; (b) measurement of surface resistivity.

(10 mHz range) symmetric voltage square wave. When testing insulators, applied voltage in the kV range is typical.

6.1.1.4   Four-point probe

The sample is a thin sheet or film of uniform thickness t. The fixture provides four contact points (for example, four spring-loaded metal tips) placed at equal distances s on a straight line, as shown in Fig. 6.4.

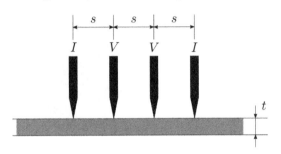

Figure 6.4
Four-point probe method, with metal tips at distance s on a material of thickness t.

For an infinite and thin $(t \ll s)$ sheet, the surface resistivity ρ_S can be calculated from the measured resistance R_S as

$$\rho_S = \frac{\pi}{\log(2)} R_S \tag{6.2}$$

For samples of finite thickness or size, correction factors to Eq. (6.2) are available (Smits, 1958).

6.1.1.5   Van der Pauw geometry

The sample is a thin sheet or film of uniform thickness t, having an arbitrary simply connected shape (no holes).

van der Pauw (1958) geometry, see Fig. 6.5, considers four point contacts A, B, C, D (in counterclockwise order) on the boundary of the sample. If 4T resistances

$$R_{AB \mapsto CD} = \frac{V_{CD}}{I_{AB}}, \qquad R_{BC \mapsto DA} = \frac{V_{BC}}{I_{DA}}$$

measured resistance R_S by the relation

$$\rho_S = R_S \frac{2\pi}{\log \frac{R_2}{R_1}}.$$

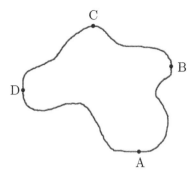

Figure 6.5
General van der Pauw geometry.

are defined, then the relation

$$\exp\left(-\pi\frac{t}{\rho}R_{AB\mapsto CD}\right) + \exp\left(-\pi\frac{t}{\rho}R_{BC\mapsto DA}\right) = 1, \tag{6.3}$$

holds. Eq. (6.3) is a strict analogue of the Thompson-Lampard theorem (Sec. 9.2.1.2).

The solution of Eq. (6.3) can be written as

$$\rho = \frac{\pi t}{\log 2}\frac{R_{AB\mapsto CD} + R_{BC\mapsto DA}}{2} \cdot f\left(\frac{R_{AB\mapsto CD}}{R_{BC\mapsto DA}}\right),$$

where the function f satisfies the relation

$$\frac{R_{AB\mapsto CD} - R_{BC\mapsto DA}}{R_{AB\mapsto CD} + R_{BC\mapsto DA}} = f \operatorname{arccosh}\left(\frac{1}{2}\exp\left(\frac{\log 2}{f}\right)\right),$$

and has to be computed numerically.

An example of a fixture suitable for applying van der Pauw method to metallic samples is shown in Fig. 6.6.

6.1.1.6 Liquids and suspensions

For liquid samples, the test fixture is called a *cell*, having known *cell constant* K (m^{-1}). In absence of any effect at the electrode-liquid interface and stray parameters, Eq. (6.1)

$$\sigma = K\,G,$$

$$\epsilon = K\frac{B}{\omega}, \tag{6.1}$$

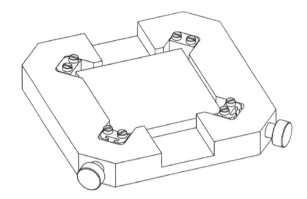

Figure 6.6
A fixture for measurement of conductivity of metallic samples, in the form of
square thick plates, with the van der Pauw method. ©2003 IEEE. Reprinted, with
permission, from Rietveld et al. (2003, Fig. 3).

still holds. The cell constant K is usually known by calibration with certified
reference materials. In cells of simple geometry the cell constant can be computed from the geometrical dimensions.[2] Different geometries give different K,
typically ranging between $0.01\,\mathrm{m}^{-1}$ and $100\,\mathrm{m}^{-1}$; the choice of the appropriate
constant is dependent on the range of σ and ϵ under investigation.

In the measurement of electrolytic solutions, at the electrode-electrolyte
interface, *polarization* phenomena occur. The *double layer* formed at the interface (a layer of ions absorbed on the electrode surface, which induces a
layer of polarization charges in the electrode) has its own impedance Z_{el}, in
series with the solution impedance Z_{s}; therefore, Eq. (6.1) is no longer valid.

Figure 6.7
Randles circuit, an equivalent circuit of the electrode-electrolyte interface
impedance Z_{el}.

A typical electrical model of Z_{el} is the *Randles circuit* (Randles, 1947), Fig.
6.7: it includes the double layer capacitance C_{dl}, a charge-transfer resistance

---

[2]For example, for a two-terminal parallel-plate cylindrical cell, having electrodes of surface area S at distance d, and where fringe field effects can be neglected or are prevented
by suitable guard electrodes, $K = dS^{-1}$.

R_{ct}, and a *Warburg impedance*, or *constant-phase element* Z_W, having the form $Z_W = A_W (j\omega)^{-\alpha}$, where A_W is a real constant and $\alpha = 0\ldots1$ ($\alpha = \frac{1}{2}$ in the original Warburg (1899) paper). On a Nyquist plot, Z_W is a oblique straight line (with $\alpha = \frac{1}{2}$, at 45°) with respect to the axes, see Fig. 6.9.

For electrolytic solutions, σ is often labeled as χ, and non-standard SI submultiple units as mS cm^{-1} and µS cm^{-1} are of widespread use.

Fig. 6.8 shows an example of a two-terminal cell for measurement of low-conductivity aqueous solution. An example of measurement result is shown in Fig. 6.9.

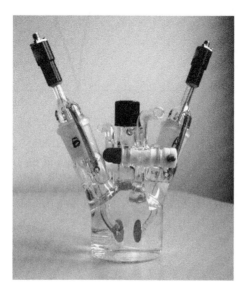

Figure 6.8
Two-terminal flow-through cell for the measurement of electrolytic conductivity of pure water. Courtesy of F. Durbiano, INRIM.

Primary cells (Brinkmann et al., 2003) are specifically designed to permit measurements that discriminate interface effects.

The design of Fig. 6.10 has a removable cylindrical central section of known geometrical dimensions (from which the section constant K is calculated). The method is differential. Two impedance measurements Z_1 and Z_2 are performed in sequence on the same sample, with the cell in two different configurations: one including the central section, giving reading Z_1, the other excluding it (Z_2). Under the hypothesis that Z_{el} is the same in both configurations, a modified version of Eq. (6.1) can be applied, $\sigma = K (R_1 - R_2)^{-1}$.

When contamination by seals or flexible elements is not an issue, primary cells where the electrode distance can be continuously varied with an external piston mechanism have been proposed (Gregory and Clarke, 2005).

For highly conductive liquids and solutions, inductive conductivity meters

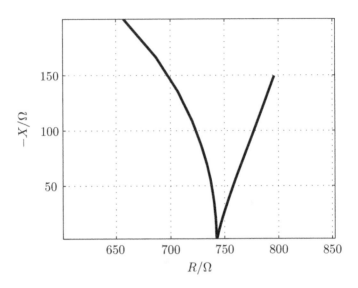

Figure 6.9
Nyquist plot $(R, -X)$ of an impedance measurement on an electrolytic solution (KCl in water, nominal conductivity $200\,\mu S$ cm^{-1}), performed with the cell of Fig. 6.8. The low-frequency region (right arm of the graph) show the linear behavior of Z_{el}, which can be modeled by a constant-phase element. Courtesy of F. Durbiano, INRIM.

Figure 6.10
Primary cell for the measurement of electrolytic conductivity of aqueous solutions, in the configuration with the cylindrical central section inserted. Courtesy of F. Durbiano, INRIM.

exist, see Fig. 6.11. The working principle (Striggow and Dankert, 1985; Waka-matsu, 1997) is to an extent similar to eddy current meters (Sec. 6.1.1.2); the liquid acts as a short-circuited secondary winding of a transformer, and its conductivity is computed from the measured impedance of the primary wind-ing. The absence of electrodes permit a construction insensitive to corrosion.

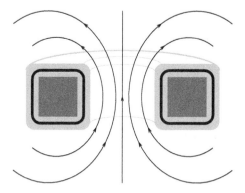

Figure 6.11
Inductive conductivity meter, simplified diagram of the measuring head. A ferromagnetic core (gray) is excited by a winding (black) and generates eddy currents in the liquid being analyzed. Adapted from Striggow and Dankert (1985, Fig. 2).

If a certain voltage threshold (either the ac voltage employed for the impedance measurement, or a dc voltage bias superimposed to it), depen-dent on the ionic species in solution, is crossed, chemical reactions at the electrodes (*ionic discharge*) occur and the ionic content evolves with time. *Electrical impedance spectroscopy* (EIS) extract information on chemical reac-tions going on by performing measurements of impedance spectra and suitable modeling (Barsoukov and Macdonald, 2005).

6.1.1.7 Gases

Measurements of gas permittivity can be conducted in a sealed gas-dielectric capacitor, by measuring its capacitance with the gas of interest C_{gas}, and then in vacuo C_{vac}. Relative gas permittivity ϵ_r is equal to the relative capacitance variation $\dfrac{C_{gas}}{C_{vac}}$, if no change in electrode spacing and surface area occurs. For more accurate measurement strain effect due to gas pressure must be accounted for, and the calculable capacitor geometry (Sec. 9.2.1.2) is more appropriate (Buckley et al., 2000; Schmidt and Moldover, 2003).

Dielectric-constant gas thermometry is a primary thermometry method, and if conducted at the triple point of water permits the determination of the Boltzmann constant (Fellmuth et al., 2011a,b).

6.1.2   Permeability measurements

The field quantities of interest for magnetic measurements satisfy the relationship

$$B(r) = \mu_0\,H(r) + \mu_0\,M(r) = \mu_0\,H(r) + J(r), \qquad (6.4)$$

where, at point r, B is the *magnetic induction* (T), H the *magnetic field* $(A\,m^{-1})$, M the *magnetization*, or magnetic moment per unit volume, $(A\,m^{-1})$, and J is the *magnetic polarization* (T).

Linear and isotropic magnetic materials satisfy the constitutive relation relationship $M(H) = \chi\,H$, where χ is the *magnetic susceptibility*, a dimensionless positive quantity. The *relative permeability* is defined as $\mu_r = 1 + \chi$, and Eq. (6.4) can be written as

$$B(r) = \mu_0\mu_r\,H(r), \qquad (6.5)$$

where μ_0 is the magnetic constant (App. C).

Diamagnetic and paramagnetic materials can be considered, up to extremely large applied field, as linear materials; therefore, susceptibility and permeability are well-defined quantities (typically, χ ranges between 10^{-6} and 10^{-3}). *Ferromagnetic* materials exhibit spontaneous magnetization and hysteresis; the constitutive law $M(H)$ becomes very complex and history-dependent, and Eq. (6.5) does not apply. The meaning of a susceptibility or permeability measurement on ferromagnetic materials must be therefore carefully considered.

The measurement techniques described in the following (see Fiorillo, 2010, for a review) are suitable when dealing with *soft* ferromagnets: materials where the condition of maximum possible magnetization (*saturation*) is achieved with a low intensity, less than $1\,kA\,m^{-1}$, of the applied field. Soft ferromagnets can be easily *demagnetized* (that is, carried to a state where $M(H = 0) \approx 0$) and, if the applied measurement field is sufficiently small (much smaller than the saturation field), Eq. (6.5) can be a reasonable approximation of their behavior.

If $M(r)$ is not uniform, a *demagnetizing field* H_d, directed against M, occurs. The phenomenon is particularly important at the sample surface, corresponding to a discontinuity in $M(r)$ (since $M \equiv 0$ in free space). Becuse of the demagnetizing field, the applied measurement field H_a is reduced within the material to an effective field $H = H_a - H_d$, of smaller amplitude. The apparent susceptibility and permeability is correspondingly reduced, and corrections must be applied to the measurement outcome. Demagnetizing effects are reduced with the choice of an appropriate sample geometry: thin films magnetized along the plane and long rods magnetized along the axis reduce demagnetizing effects, and in closed magnetic circuits (frames and rings) demagnetizing effects can be neglected.

In the sinusoidal steady state, the quantities $\boldsymbol{B}(\boldsymbol{r})$ and $\boldsymbol{H}(\boldsymbol{r})$ of Eq. (6.5) can be reconsidered as phasors, related by complex quantities $\chi = \chi' - j\chi''$ and $\mu'_r - j\mu''_r$, where $\mu'_r = \chi' + 1$ and $\mu''_r = \chi''$ are real quantities, permit to express also the magnetic energy loss.[3] Consider an inductor of n turns uniformly wound on a magnetic circuit having effective length ℓ and effective cross section S.[4] The magnetic flux Φ in the magnetic circuit due to current I can be written as

$$\Phi = B S = \mu_0 \mu_r H S = \mu_0 \mu_r \frac{nI}{\ell} S,$$

and the electromotive force in the winding is

$$E = n j\omega \Phi = j\omega \mu_0 \mu_r \frac{n^2 S}{\ell} I.$$

If the winding has resistance R_w, the voltage on the inductor is $V = R_w I + E$, giving an impedance

$$Z = R_w + \frac{E}{I} = R_w + j\omega \mu_0 \mu_r \frac{n^2 S}{\ell}, \tag{6.6}$$

which equivalent series representation is

$$R = \mathrm{Re}[Z] = R_w + \omega\mu_0 \mu''_r \frac{n^2 S}{\ell},$$

$$X = \mathrm{Im}[Z] = \omega\mu_0 \mu'_r \frac{n^2 S}{\ell}.$$

Therefore, the complex permeability value can be derived by an impedance measurement on a suitable fixture.

Closed magnetic circuits suitable for permeability measurements can be formed in different ways, see Fig. 6.12.

In the audio frequency range, the measurement can performed by constructing a self- or a mutual inductor with the closed magnetic circuit as core. At higher frequencies, a more suitable arrangement is a coaxial shorted line (Sec. 6.1.3), acting as a one-turn self-inductor (Goldfarb and Bussey, 1987).

---

[3]The minus sign in the definition of the complex value of μ_r and χ is there to have $\mu''_r = \chi'' \geq 0$ in real materials.

[4]The relation of effective quantities ℓ and S with geometrical dimensions is dependent on sample geometry. For a long straight bar, neglecting the demagnetizing field, ℓ and S correspond to the geometrical dimensions. For a toroid of rectangular cross-section, having inner radius r, outer radius R, and height h,

$$\ell = 2\pi \frac{R+r}{2}, \qquad S = h \frac{R+r}{2} \log \frac{R}{r}.$$

Figure 6.12
Methods to employ magnetic materials to achieve closed magnetic circuits.
From left to right: by assembling strips on a frame (Epstein frame), with
a detail on how the strips are stacked; by stacking rings punched out of a
lamination; by winding a ribbon; by making a solid ring with sintered/bonded
powders. Courtesy of F. Fiorillo, INRIM.

6.1.3   Transmission line measurements

The electromagnetic properties of the material employed as dielectric in
a transmission line (Sec. 1.6.4.4) have a direct influence on its electrical
properties. A measurement of the transmission line properties, and knowledge of its geometrical properties, permit to recover the complex permittiv-
ity/permeability of the material (Baker-Jarvis et al., 1993). Coaxial transmission lines, or waveguides, are particularly suited for the purpose, since the
material can be fully enclosed in the (outer) conductor; therefore, both solid
and liquid materials can be conveniently measured. As examples, Fig. 6.13
shows a 1P fixture for dielectric liquid measurements; Fig. 6.14 shows a 1P
shorted line for the measurement of toroidal-shaped magnetic materials.

Transmission line measurements are often performed in a wide frequency
range, with network analysis (Sec. 4.11), either with more conventional signal
frequency sweeps or with TDR (Sec. 4.11.3) (Fellner-Feldegg, 1969; Feldman
et al., 1996; Pettinelli et al., 2002).

6.1.4   Impedance tomography

In *electrical impedance tomography* or EIT (Holder, 2005), images of the object are formed on the basis of the local volume electromagnetic properties
(conductivity and permittivity). A large number of four-terminal impedance
measurements are performed by connecting the meter(s) to electrodes placed
on the external surface of the object under investigation. Fixed-frequency or
spectroscopical methods can be employed. Results of such measurements enter
a mathematical model that reconstructs the inner volume local conductivity
and permittivity, which is then associated to local variations in composition,
density, and phase state. Major applications of EIT are in imaging of multi-
phase flow in pipes, and in vivo biological measurements (e.g., of the human
chest).

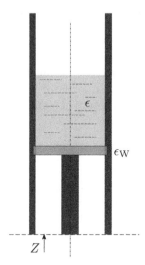

Figure 6.13
Coaxial line to circular waveguide transition employed as a fixture for dielectric liquids. 1P impedance measurement permit, with suitable modeling, to determine the permittivity ϵ of the liquid poured in the waveguide. The transmission line includes a solid dielectric window (with permittivity ϵ_w) that confines the liquid. Adapted from Kaatze and Feldman (2006, Fig. 9).

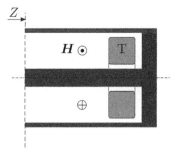

Figure 6.14
Shorted coaxial transmission line for the measurement of permeability. The toroidal sample T is inserted in the line and excited by the magnetic field H developed by the measuring current; the impedance Z is measured at the reference plane.

6.2 Sensor reading

If a material changes its electromagnetic properties because of changes of a physical quantity, a chemical substance, or a biological agent, and one of these properties (resistance, permittivity, permeability) is sensed with one of the techniques given in Sec. 6.1, then a sensor having impedance as its output is, in principle, obtained. A few sensors of common use are listed in this section.

6.2.1 Resistance thermometers

Resistive thermometers (platinum resistance thermometers, thermistors, low-temperature semiconductor sensors) are widespread. Resistance measurement in the ac regime permits to achieve higher measurement sensitivity, and thus reduce the electrical excitation of the sensor (which cause self-heating), and to reject thermoelectromotive forces caused by temperature differences in the wiring.

Standard platinum resistance thermometry (Strouse, 2008) achieves measurement accuracies of $1\,\mathrm{mK}$ or better. Since the temperature coefficient of platinum resistivity is $\approx 4 \times 10^{-3}\,\mathrm{K}^{-1}$ at room temperature, the corresponding relative resistance meter accuracy to achieve such accuracy has to be in the 10^{-6} range or better. Commercial automated bridges, based either on inductive voltage dividers or on current comparators, can measure 4C resistance ratios at low frequency (typically $1/2$ or $3/2$ of mains frequency, e.g., $25\,\mathrm{Hz}$ or $75\,\mathrm{Hz}$ for a $50\,\mathrm{Hz}$ mains, to avoid interferences).

6.2.2 Capacitive hygrometers

The sensing material of a capacitive hygrometer is a hygroscopic isolating material (a polymer, or a metal-oxide ceramics). The material constitutes the dielectric of a capacitor C; the electrode surfaces are made porous to permit continuous contact with external air.

The permittivity of the material, and hence C, increases with the relative humidity of air; the response is approximately linear between $5\,\%$ RH and $95\,\%$ RH. Typical C is in the $100\,\mathrm{pF}$ range, with a relative variation of C over full-scale humidity change of $0.2\,C$.

6.2.3 Displacement sensors

6.2.3.1 Capacitive sensors

The dependence of a gas-dielectric capacitance C on the geometry, in particular on the distance d between the electrode plates, can be employed for non-contact displacement measurement.

Capacitive sensors can achieve sub-nm sensitivity and are employed in high-precision machining (Small and Fiander, 2011) and positioning (Cabiati and D'Elia, 2000; Picotto and Pisani, 2001). Ultimate-sensitivity capacitive sensors are employed in gravitational detectors (Gottardi et al., 2007). Micromachined pressure sensors can employ capacitive reading of a diapraghm displacement (Eaton and Smith, 1997).

Since capacitive sensors have often a small capacitance value (typically in the range 1 pF to 100 pF), achieving high sensitivity and accuracy require very sensitive meters and appropriate shielding from electromagnetic interferences.

6.2.3.2 Capacitive vibration sensors and microphones

Capacitance vibration sensors and microphone capsules are based on an extension of the capacitance definition (1.2.2) to time-dependent capacitance: $q(t) = C(t)v(t)$, which time derivative is $i(t) = C\dfrac{\partial v}{\partial t} + v(t)\dfrac{\partial C}{\partial t}$.

If capacitor C is polarized at dc voltage $v(t) = V$ and vibrates at angular frequency ω, in the linear approximation one can write $C = C_0 + K\cos(\omega t)$; the resulting current output $i(t) = -KV\omega\sin(\omega t)$ is proportional to the vibration amplitude K, and to frequency ω.

Transconductance (Sec. 3.2.2) or charge amplifiers can be employed for the measurement. Because of the high output impedance of the sensor (its capacitance is usually of a few pF), the current output $i(t)$ is shunted by the capacitance of the connections and the amplifier input. To avoid sensitivity loss, triaxial connections and active guarding (Sec. 3.5.1) is often employed.

6.2.3.3 Inductive sensors

Inductive sensors rely on inductance L variations when the magnetic flux path is modified by the presence of a conducting or ferromagnetic body nearby. In most applications, the working principle is similar to eddy-current resistance measurement (Sec. 6.1.1.2), but the quantity which is determined is not the resistivity but the distance of the conducting plate from the sensing coil. Most applications of inductive sensors are as binary-output *proximity sensors*.

6.2.4 Magnetic field sensors

The measurement of an ac magnetic flux Φ can be performed with an inductor, where an electromotive force $E = -j\omega\Phi$ is induced. Typical geometries are multi-turn circular coil in air.

6.2.4.1 Fluxgates

The *fluxgate* magnetometer does not involve an impedance measurement, but is here cited because the same detection principle is employed in magnetic current comparators for dc currents (DCC), Sec. 4.5.1.

The device (Primdahl, 1970, 1979; Ripka, 2001) is based on an open ferromagnetic core, exposed to the environmental dc magnetic field to be measured (which can be considered as source of dc magnetomotive force M_{dc}). A flux Φ is generated in the core. In DCCs, the flux Φ is generated by the unbalanced M_{dc} given by primary windings' currents.

An excitation winding driven at audio frequency excites the core with a sinusoidal ac magnetomotive force $M_{ac}(t)$. The resulting magnetic flux in the core is a function $\Phi(t) = f(M(t))$ of the total applied magnetomotive force $M(t) = M_{dc} + M_{ac}(t)$. The flux $\Phi(t)$ can be sensed as the electromotive force $E(t)$ on a detection winding.

Since, in a ferromagnetic core, $f(M)$ is nonlinear but antisymmetric (that is, the relation $f(-M) = -f(-M)$ holds), it follows that if $M_{dc} \equiv 0$ then $\Phi(t)$ (and, consequently, $E(t)$) contains, as in ac transformers, only odd harmonics. When instead $M_{dc} \neq 0$, $\Phi(t)$ (and $E(t)$) includes also even harmonics. The second harmonic of $E(t)$ can be detected by a suitable demodulation electronics.

Feedback fluxgates, and all DCCs, are operated at zero flux by nulling M_{dc} with a compensation winding driven by a dc current.

6.2.4.2   Magnetoimpedance sensors

Magnetoimpedance sensors (Hauser et al., 2001; Ripka, 2001) are sensors of external dc magnetic field H; the working principle is the dependence of the measured impedance Z of soft ferromagnetic wires on H. Magnetoimpedance effect is caused by the dependence of the permeability $\mu(H)$ of the wire with H, which in turn causes a variation of the penetration depth δ (Sec. 1.10.1) of the electric current flowing in the wire. For a circular and magnetically isotropic wire of radius a, the wire impedance can be written as (Menard and Yelon, 2000, Eq. 34)

$$\frac{Z}{R_{dc}} = \frac{ka}{2} \frac{J_0(ka)}{J_1(ka)}, \qquad k = \frac{1-j}{\delta}, \tag{6.7}$$

where R_{dc} is the dc resistance of the wire, $J_0(x)$ and $J_1(x)$ are the zero- and one-order Bessel functions of first kind. If the measurement frequency ω is sufficiently high such that $\delta \ll a$, Eq. (6.7) can be rewritten as

$$\frac{Z(H)}{R_{dc}} = (1+j)\frac{a}{2\sqrt{2}\rho} R_{dc} \sqrt{\omega\mu(H)}. \tag{6.8}$$

Although in real devices deviations from Eq. (6.8) occur, even such an ideal case shows that the sensitivity increases with ω, which in practical devices is in the range $10\,\text{kHz}$–$20\,\text{MHz}$. Sensor reading is typically performed by the resonance method (Sec. 4.9) or in an unbalanced Wheatstone bridge.

6.3 Semiconductor device characterization

Impedance measurements are involved in the characterization of semiconductor devices. Two major characterization techniques are the following.

6.3.1 Capacitance-voltage (C-V)

In *capacitance-voltage* (C-V) characterization, the semiconductor junction of interest is polarized with a dc bias V_{dc}; the junction impedance is measured with a small ac voltage signal (typical frequency range 100 kHz–1 MHz) superimposed to V_{dc}. Capacitance (and loss) is plotted as function of V_{dc}.

As example of application of C-V method, consider a PN junction diode, reversely polarized by V_{dc}. The diode capacitance is well approximated by the depletion region capacitance, which width W is dependent on V_{dc}. In the parallel-plate approximation, if S is the junction area and ϵ the semiconductor permittivity,

$$C(V_{dc}) = \epsilon \frac{S}{W(V_{dc})}. \tag{6.9}$$

It can be shown (Sze and Ng, 2007, p. 85) that the doping concentration $N(W)$ for a given junction can be related to $C(V_{dc})$ by

$$N(W) = -\frac{2}{\epsilon \, S^2 \, q} \cdot \left[\frac{\mathrm{d}C^{-2}(V_{dc})}{\mathrm{d}V_{dc}} \right]^{-1}, \tag{6.10}$$

where q is the elementary charge. Eq. (6.9) and (6.10) combined permit to obtain the doping profile $N(W)$ from $C(V_{dc})$.

More recently (Straub et al., 2005) an extension of the C-V method called *impedance analysis* has been proposed, where bias-dependent impedance spectroscopy $Z(f; V_{dc})$ is performed on the device, and the large amount of experimental data employed to identify the a more complete electrical model than that given by Eq. (6.9) and (6.10).

6.3.2 Deep-level transient spectroscopy (DLTS)

Deep-level transient spectroscopy (DLTS) (Lang, 1974) is an experimental techinque for the characterization of electrical defects, known as *carrier traps*, in a semiconductor.

The technique relies on the measurement of the time evolution of the capacitance $C(t)$ under an abrupt change of the polarization voltage bias $v(t)$; the measurement is typically performed at several different temperatures. For example, a PN junction, reversely polarized ($v(t) = -V_{bias}$) can be unpolarized ($v(t) = 0$) for a short time. Traps are filled by charge carriers during the depolarization pulse, and then slowly[5] release trapped charges, modifying the resulting electric field profile and hence the junction capacitance. The analysis of relaxation capacitance curves permits to obtain quantitative information on the traps. In silicon, the sensitivity to impurity concentration can exceed 10^{-12}.

6.4   Biological measurements

Impedance measurements on biological materials, both *in vitro* and *in vivo*, is a quickly growing subject in literature, quite difficult to summarize briefly here. The typical measurement is spectroscopic (Sec. 5.5.1). Measurements can be performed on the living body (Oldham, 1996; Kyle et al., 2004a,b); on bodily fluids, such as blood (Bao et al., 1993); on cellular cultures (Berggren et al., 2001; Yang et al., 2004); on genetic material (Long et al., 2003).

---

[5]Typical time constants are strongly temperature dependent and can range from 10^{-4} to 10^4 s.

7

Metrology: traceability and uncertainty

CONTENTS

7.1   Traceability

7.1.1   General definition of calibration and traceability

The VIM defines a number of metrology terms employed throughout this book:[1]

Calibration is the "operation that, under specified conditions, in a first step, establishes a relation between the quantity values with measurement uncertainties provided by measurement standards and corresponding indications with associated measurement uncertainties and, in a second step, uses this information to establish a relation for obtaining a measurement result from an indication" (VIM, 2.39).

Adjustment is the "set of operations carried out on a measuring system so that it provides prescribed indications corresponding to given values of a quantity to be measured" (VIM, 3.11).

Metrological traceability is the "property of a measurement result whereby the result can be related to a reference through a documented unbroken chain of calibrations, each contributing to the measurement uncertainty" (VIM, 2.41). "For this definition, a 'reference' can be a definition of a measurement unit through its practical realization, or a measurement

---

[1]Excerpts from VIM are published with permission of the BIPM director. The BIPM director is the chairman of the JCGM. The JCGM does not accept any liability for the relevance, accuracy, completeness, or quality of reproduced information and materials. The only official version is the original version of the documents published by the JCGM.

procedure including the measurement unit for a non-ordinal quantity, or a measurement standard" (VIM, 2.41, note 1). "Metrological traceability requires an established calibration hierarchy" (VIM, 2.41, note 2). "For measurements with more than one input quantity in the measurement model, each of the input quantity values should itself be metrologically traceable and the calibration hierarchy involved may form a branched structure or a network. The effort involved in establishing metrological traceability for each input quantity value should be commensurate with its relative contribution to the measurement result" (VIM, 2.41, note 4). "A comparison between two measurement standards may be viewed as a calibration if the comparison is used to check and, if necessary, correct the quantity value and measurement uncertainty attributed to one of the measurement standards" (VIM, 2.41, note 6).

The benefit of traceability in measurement is the equivalence of measurement outcomes made independently of one another. If two measurements of a quantity are made independently, and these measurements are both traceable, then their values will agree to within the combined stated uncertainties.

7.1.2 Traceability in impedance measurement

All measurement methods described in Ch. 4 and 5 perform a comparison between the impedance to be measured and one or more reference impedance standards. In the *I-V* method (Sec. 4.1) the standard is embedded in the current-to-voltage converter. In impedance bridges (Sec. 4.4) the standard is an independent artifact connected to the bridge circuitry by the user. In LCR meters (Sec. 4.7) the standards are internal components of the meter, selectable by internal switches but not accessible from the outside. In network analyzers (Sec. 4.11) the standard is the characteristic impedance Z_0 of the analyzer components and connections.

Traceability in impedance measurement can be achieved in several ways, not mutually exclusive:

- by realization or reproduction of the impedance unit (Ch. 9), where the impedance value is given by non-electrical measurements or is linked to fundamental constants of nature.

- by calculable standards, where the quantity of interest (e.g., frequency dependence) is derived from non-electrical measurements (e.g., calculable resistors, Sec. 8.1.2.6).

- by separate calibration of the impedance standard(s) employed in the measurement circuit, and of all other components with which properties enter the measurement model (e.g., for a voltage transformer ratio bridge, the voltage transformer ratio under operating conditions). It is possible that some components do not require a calibration but simply a preliminary adjustment (e.g., the offset compensation of a detector).

- by "calibration" of the whole measuring circuit, more properly an adjustment, with a set of calibrated impedance standards. Examples include the calibration of LCR meters (Sec. 4.7), and of network analyzers and six-port reflectometers (Sec. 4.11). Usually, in performing the calibration, some properties of the whole meter (or of the individual components involved) are assumed (e.g., linearity); these properties should be verified with independent measurements.

- by direct reading with a calibrated instrument. Most impedance meters have a lower stability (versus time and environmental conditions) than impedance standards for a corresponding impedance magnitude and frequency of interest; it follows that, respect to previous listed methods, a lower measurement accuracy may be achieved.

In primary metrology a traceability chain starts with the realization or reproduction of an impedance unit; well-established ways are given by the dc quantum Hall effect (Sec. 9.3.1) and the calculable capacitor (Sec. 9.2.1.2). The chain then logically[2] proceeds by comparisons performed with dedicated measurement setups, until the calibration of a set of impedances (maintained standard) of very high stability is achieved. Fig. 7.1 shows an example of traceability chain for the realization of the capacitance unit from the dc quantum Hall effect.

7.2 Uncertainty

7.2.1 Statements from VIM

The goal of *measurement* (VIM, 2.1) is to obtain experimentally a *measurement result* (VIM, 2.9), the set of quantity values, together with any other available relevant information, attributed to a *measurand* (VIM, 2.3). A measurement result is generally expressed as a single measured quantity value and a *measurement uncertainty* (VIM, 2.26). The measurement uncertainty is a parameter that characterizes the dispersion of quantity values being attributed to a measurand. The parameter can be the *standard measurement uncertainty* or an interval having a stated coverage probability, called *interval of confidence*. Measurement uncertainty comprises in general many components (VIM, 2.26, note 3). Some of these may be evaluated by *Type A evaluation of measurement uncertainty* from the statistical distribution of quantity values from a series of measurements, and can be characterized by standard deviations. The other components are evaluated by *Type B evaluation of measurement uncertainty* from probability density functions based on experience or other information

---

[2]The ordering in time of the comparisons can be different from the logical ordering implied in the traceability chain.

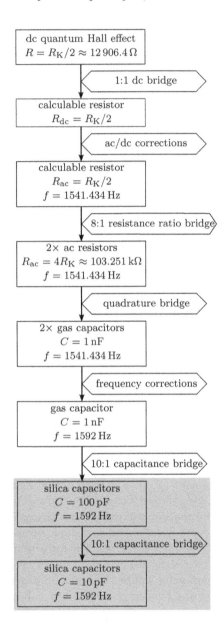

Figure 7.1
A traceability chain to realize the capacitance unit (the gray area is the maintained standard, at 10 pF and 100 pF level, with highly stable fused silica capacitors, Sec. 8.3.2) from the dc quantum Hall effect. All values reported are nominal values. The particular frequency values have the following meaning: $f = 1592\,\mathrm{Hz}$ corresponds to $\omega \approx 10\,\mathrm{krad\,s^{-1}}$; at frequency $f = 1541.434\,\mathrm{Hz}$, a 1 nF capacitor has a reactance equal to $4R_\mathrm{K}$, to permit comparison with the quadrature bridge. Both the resistance ratio bridge and the quadrature bridge are digitally assisted bridges (Sec. 5.4.1). Details are given in Callegaro et al.

available: for example, associated with authoritative published quantity values, obtained from a *calibration certificate* or from instrument specifications (VIM, 2.29).

The *measurement model* (VIM, 2.48) is a mathematical relation among all quantities known to be involved in a measurement. Such quantities are the *input quantities* (VIM, 2.50) (quantity values measured or obtained otherwise) including *instrument indications* (VIM, 4.1), *influence quantities* (VIM, 2.52), and the output quantity, the measurand value.

7.2.2 Impedance measurement and uncertainty expression

In previous chapters a number of measurement methods and techniques of impedance measurement have been reviewed. Addressing the problem of uncertainty evaluation, even for a subset of the methods reported, would have required excessive space. Usually, together with the description of each method, several references to research papers are given; most papers deal in some way with the uncertainty of reported data, and a number provide also a detailed budget of the various contributions to uncertainty. What follows in this section is therefore just a list of general references.

The fundamental reference document of uncertainty expression is the *Guide to the Expression of Uncertainty in Measurement* (GUM). The GUM addressses univariate measurement models, namely, models having a single scalar (real-valued) output quantity.

In electrical impedance metrology, a measurement model is often expressed with complex quantities; this can include the measurand itself. Whereas the VIM definitions can be readily extended to include vector and complex-valued input quantities and measurands, the uncertainty evaluation ask for a nontrivial approach. This is the subject of the *Supplement 2 to the GUM – Extension to any number of output quantities* (GUM Suppl. 2). Both GUM and GUM Suppl. 2 calculations can be performed analytically, or by numerical calculations: this is the subject of *Supplement 1 to the GUM –Propagation of distributions using a Monte Carlo method* (GUM Suppl. 1).

Presently, the subject of uncertainty expression in impedance measurements is a research field, since many problems are still open.

From a general point of view, the framework of GUM Suppl. 2 does not permit a direct treatment of integral relations: therefore, for example, frequency correlations given by Kramers-Krönig relations (Sec. 1.7.4.1) cannot be properly taken into account; the application of GUM Suppl. 2 to digital signal processing (Sec. 5.3), in principle possible, is extremely cumbersome.

From the point of view of impedance measurement practice, achieving a proper measurement model (the necessary step for uncertainty expression) for a given instrument or measurement setup asks for a detailed physical and electrical modeling; and, when commercial instruments are employed, a deep understanding of the – ever increasing – complexity of the digital algorithms embedded in the firmware.

8

Metrology: standards

CONTENTS

8.1 Impedance standards

8.1.1 Construction goals

The construction of artifact standards of impedances aim to reach some goals. Some of the properties that define an impedance standard are listed below.

accuracy class The impedance value of the standard (as given by a calibration) should not be too far from its *nominal value*, determined by its construction. The nominal value can be printed on the standard itself, or determined (for decade boxes) by the dial settings, see Fig. 8.1. The *accuracy class* of a standard is typically expressed as the maximum permissible deviation of the true value with respect to the nominal value, in percentage.

(a)   (b)

Figure 8.1
(a) The nominal value of a Sullivan & Griffiths 100 μH mutual inductor is very clearly stated on its label, here shown. (b) With such dial setting, the nominal value of this General Radio mod. 1419-K capacitance box is 0.56 μF.

type of definition The natural electrical definition of the impedance standard given by the number and type of its connections. For example, a 4P standard will have four coaxial connector on its body. See Ch. 2.

connectors' quality The electrical connections introduce parasitic effects (contact resistances, thermoelectromotive forces, stray capacitance and inductance, stray conductances). A good connection permits to minimize such effects, and achieve the maximum reproducibility between connections. The connection quality is particularly relevant for simpler definitions (2T, 3T, 2P), see Ch. 2.

time stability The drift of the standard value with time should be as small as possible. Residual unavoidable drifts should have a smooth behavior with time, avoiding unexpected jumps, and therefore be predictable in time between calibrations.

environmental coefficients The dependence of the standard value on environmental temperature, humidity, and air pressure should be as small as possible. The dependencies can have very long time constants (even weeks for humidity) or hysteresis.

measurement condition dependencies The dependence of the standard value on the properties of the electrical excitation employed during measurement; a low dependence permit to relax the accuracy required in reproducing the excitation on different measurements or instruments. The dependence is often stated in terms of dependence coefficients: examples are *voltage* or *current coefficient, power coefficient*, either specified for the standard type, or measured on the particular item. The maximum allowed excitation value is typically specified.

purity and frequency dependence A *pure* impedance is a resistor with zero time constant, a capacitor with zero dissipation factor, or an inductor with infinite quality factor; the inevitable deviations from purity should be as small as possible. The value of the primary parameter (e.g., capacitance for a capacitor) should have the lowest dispersion versus frequency as possible. Purity and frequency dependence are interrelated (Sec 1.7.4.1).

8.1.2 Resistance standards

Resistance standards of interest for impedance metrology are typically realized in two ways: wound metal wire/tape, or thick metal film. Various metal alloys ar chosen for construction: some brand names are manganin ($Cu_{0.86} Mn_{0.12} Ni_{0.02}$), Evanohm ($Ni_{0.75} Cr_{0.20} Al_{0.025} Cu_{0.025}$), Karma ($Ni_{0.74}, Cr_{0.20} Fe_{0.03} Al_{0.03} Si Mn C$), Isaohm ($Ni_{0.745} Cr_{20} Al_{0.035} Si_{0.01} Mn Fe$).[1] For many alloys, delicate heat treatments permit not only to reduce residual mechanical stresses in the alloys but also to achieve a minimal temperature coefficient at the working temperature. Other materials, if adequate for the construction of electrical components of common use (such as semimetal or semiconductive materials, electrolytes etc), usually have too large temperature coefficients and drift to be adequate for standard construction, and their use is restricted to very special cases (e.g., semiconductive materials for very high resistance values).

8.1.2.1 Wire resistors

Wirewound resistance standards are realized with resistive metallic alloys on insulating supports. The winding shape is chosen to minimize the resulting parasitic parameters. To minimize inductance, the current path should embrace the smallest possible area; to minimize capacitance, nearby sections of the wire should have similar electrical potential. The wire strain is minimized by a careful winding and a thermal annealing, since stresses in the wire can be abruptly released and cause resistance jumps during the resistor life.

The construction of resistance standards having a "good" frequency performance (that is, low frequency dependence and small phase angle) is the

---

[1] The alloy compositions are approximate. All brand names are registered trademarks.

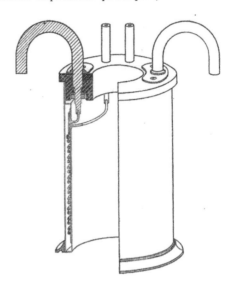

Figure 8.2
Drawing of the construction of a cylindrical resistance standard, showing the
bifilar winding in a double-walled container. After Thomas (1930, fig. 2). Official
contribution of NIST; not subject to copyright in the US.

subject of a large number of papers since the end of 19th century: a review
of several construction methods can be found in Hague (1971, Ch. 2). Fig.
8.2 shows a typical wire resistance standard construction, where the bifilar
winding minimizes the inductance. The Wilkins and Swan (1970) construction, see Fig. 8.3, permits a further reduction of the resistor time constant
and resistance frequency dependence. The winding can be also made on a flat
mica former, as shown in Fig. 8.4.

8.1.2.2   Film resistors

In film resistors, a meandering resistive path is obtained by optical lithography
on a planar metal film deposited on an insulating substrate, see Fig. 8.5(a).

The resistor is bonded to two or four terminals and encased in a plastic
or metal package, as shown in Fig. 8.5(b). Because of the geometry and the
small physical dimension, both parasitic inductance and capacitance are very
small.

8.1.2.3   High-resistance standards

High-valued resistor components, if suitable for the construction of dc resistance standards, have a parasitic parallel capacitance (typically in the 100 fF

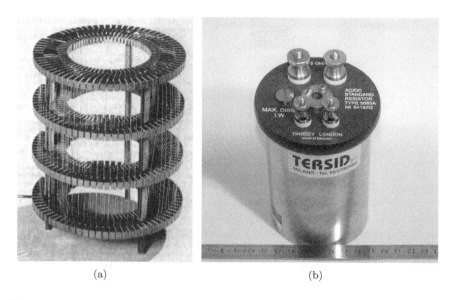

(a)   (b)

Figure 8.3
(a) The Wilkins ac/dc standard resistor wiring arrangement. Reproduced with permission of IEE (United Kingdom), from Wilkins and Swan (1970, Fig. 2); permission conveyed through Copyright Clearance Center, Inc. (b) A Tinsley 5685A 25 Ω ac/dc standard resistor, employing Wilkins geometry.

Figure 8.4
Electro Scientific Industries (ESI) 100 kΩ two-terminal wirewound resistor.

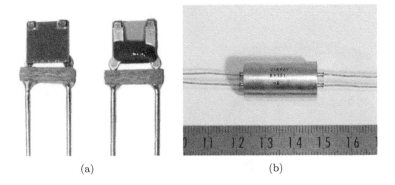

(a)   (b)

Figure 8.5
(a) A 2T film resistor, before being encased in molded plastic. (b) A Vishay mod. VHA512 4T metal foil resistor, encased in a metal oil-filled package; custom-made to have a nominal value of $6.4532\,\mathrm{k\Omega} \approx R_K/4$, where R_K is the von Klitzing constant (Sec. 9.3.1). The deviation from nominal value is less than 1×10^{-5}, the temperature coefficient is better than $2 \times 10^{-6}\,\mathrm{K^{-1}}$.

range), which give unacceptably large time constants in the ac regime. Further, unexpected resistance frequency dependence can occur (Price, 2002).

For these reasons, high-valued ($>100\,\mathrm{M\Omega}$) 3T or 2P ac resistance standards are constructed with three-resistance networks as shown in Fig. 8.6. A disadvantage of the network is the much higher current thermal noise than one could expect from its nominal resistance value.

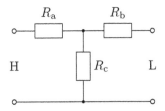

Figure 8.6
2P resistance network simulating a high-resistance value. The equivalent 2P resistance is $R = (R_a R_b + R_b R_c + R_a R_c)/R_c$.

8.1.2.4   Shunts

Low-resistance ac standards, or *shunts*, are characterized by large size of body and connectors, to handle high current values. They are defined as 4T or 4C impedance standards. The geometry of current paths is chosen to minimize the generated magnetic flux, which in turn minimize the inductance and time constant of the standard. An example of shunt is shown in Fig. 8.7.

Figure 8.7
Coaxial shunt, open view. The resistive element is a thick-film disk resistor on alumina substrate. INRIM custom realization, 1 Ω nominal value. Courtesy of U. Pogliano, INRIM.

8.1.2.5 Decade resistance boxes

Resistance boxes are variable resistors consisting of a case (electrostatic screen) containing a number of resistance components; the number of resistors in series, and consequently the resulting resistance value at the connection terminals, can be varied by acting on switches. In decade boxes (Fig. 8.8), there is a rotary switch for each resistance value digit. The resistance box is typically defined as a 3T standard with binding posts. When all decades are set to 0, the resistance box has still a minimal resistance value (the series of all internal connections and contact resistance of all switches). The frequency dependence and the time constant of a resistance box is usually much larger than that of a fixed resistor in the same value range.

8.1.2.6 Calculable resistors

When the geometry of a resistor is particularly simple and symmetric, its frequency performance can be computed on the basis of its dimensions and of the properties of materials employed (resistivity of resistive material and of conducting shields, permittivity of the dielectric). The main geometries of interest are:

coaxial is the simplest geometry: a straight resistive element with cylindrical symmetry (typically, a bare wire) is suspended coaxially inside a cylin-

Figure 8.8
6-decade resistance box General Radio mod. GR1433-W, $10\,\text{m}\Omega$ to $11.1111\,\text{k}\Omega$ with $10\,\text{m}\Omega$ resolution.

drical shield, see Fig. 8.9. The element is contacted at its extremities to provide the appropriate impedance definition; the basis for a calculation is electrical coaxiality (that is, the shield carries a return current with same amplitude and opposite direction of the wire). Experimental realizations and partial calculations are disseminated in several papers (see, e.g., Astbury (1935); Wilkins and Swan (1969)); the first complete calculation and four terminal-pair realization is given by Haddad (1969). Examples of modern realization for the audio frequency range are described in Elmquist (2000), Gulmez et al. (2002), and Kucera et al. (2009).

Haddad design has two drawbacks: the resistor connections are on two different planes (which usually require long cables); the electrical coaxiality (Sec. 3.5.2) between the inner wire element and the outer shield, on which the calculation is based, has to be strictly ensured by the external measuring circuitry. Modified designs have been proposed (Awan and Kibble, 2005; Kucera et al., 2009) to solve these drawbacks, see Figs 8.10 and 8.11: the connections are placed on the same plane, and a double-shielded construction is employed, resembling a four terminal-pair version of Wilkins and Swan (1969) design.

Calculable thin film resistors have also been realized (Bounouh, 2004, 2006; Bounouh et al., 2008), see Fig. 8.12. High resistive values can be obtained with short resistive lengths (up to $100\,\text{k}\Omega$ with a length of a few cm), giving a very small frequency dependence and time constant.

bifilar, quadrifilar, octofilar; a long resistive element can be folded in two (bifilar), four (quadrifilar), eight (octofilar), see Fig. 8.13; the purpose is to have both terminals on the same side of the resistor (in such way that the shield does not carry the main current) and to reduce the overall length of the resistor. The arrangement is still highly symmetric; exact expressions for resistance dependence with frequency have been worked out by Gibbings (1963) for bifilar and quadrifilars, and by Bohacek and

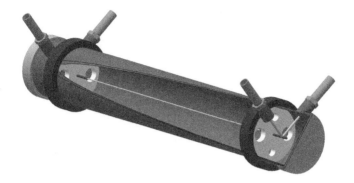

Figure 8.9
Haddad calculable resistor, cutaway drawing. The standard is defined as a
4TP impedance: the resistance wire is supported at the axis of the cylindrical
shield by PTFE disks, and connected at each end to two BPO connectors.
Courtesy of J. Kučera, CMI.

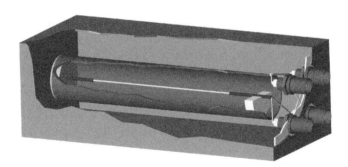

Figure 8.10
Modified Haddad calculable resistor (Kucera et al., 2009), cutaway drawing.
At variance with the design of Fig. 8.9, the cylindrical shield is now in series with the resistance wire, and an external electrostatic shield is added;
all connectors are on the same plane. For a resistor with a nominal value of
$1\,\mathrm{k\Omega}$ constructed with an Isaohm wire of $20\,\mathrm{\mu m}$ diameter, the wire length of
$\approx 350\,\mathrm{mm}$. The calculated (up to $30\,\mathrm{kHz}$) frequency dependence for such resistor is approximately parabolic, $\Delta R/R = Kf^2$, with $K = 1.6 \times 10^{-10}\,\mathrm{kHz}^{-2}$.
Courtesy of J. Kučera, CMI.

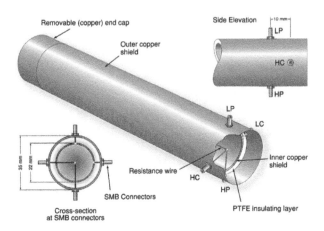

Figure 8.11
Modified Haddad calculable resistor, suitable for high-frequency measurements. The NPL design has been employed for the construction of two resistance standards having nominal values of $100\,\Omega$ and $1\,\mathrm{k}\Omega$; the relative resistance variation at $1\,\mathrm{MHz}$ is $30\,\mu\Omega\,\Omega^{-1}$ and $-40\,\mu\Omega\,\Omega^{-1}$ respectively. ©2005 IEEE. Reprinted, with permission, from Awan and Kibble (2005, Fig. 4).

Wood (2001) for the octofilar design. The most common arrangement for audio frequency is the quadrifilar, extensively employed in the realization of capacitance unit from the quantum Hall resistance. The octofilar shows a lower frequency dependence than quadrifilar designs of similar physical size (Bohacek and Wood, 2001). In the higher-resistance range, a more compact arrangement can be obtained if the bare wire is replaced with a glass-supported microwire (Semyonov et al., 1997); however, dielectric losses in the supporting glass structure should be taken into account in the calculations, hence an adequate knowledge of deposited glass electrical properties is needed.

Bifilar realizations dedicated to high-frequency measurements, covering the $1\,\Omega$-$100\,\mathrm{k}\Omega$ range have been realized (Kim et al., 2006) and tested up to $1\,\mathrm{MHz}$.

8.2 Inductance standards

Inductance standards are made of copper windings on a core. A ferromagnetic core permits to obtain high inductance values, with high Q, with a relatively

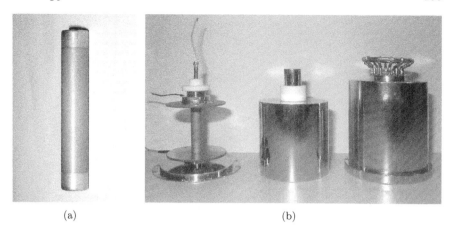

(a)   (b)

Figure 8.12
A thin-film calculable resistor. (a) The ceramic stick with resistive NiCr layer
and gold contacts. (b) The mounting assembly. Current contacts are provided
by clamps on the stick, the return current path is through the outer screen.
Voltage contacts are given by gold-plated glass disks and an inner screen.
For $1\,k\Omega$ resistor, finite-element calculations give a resistance frequency de-
pendence of 5×10^{-9} at $1\,kHz$, and a time constant of $-120\,ps$. Courtesy of A.
Bounouh, LNE.

small number of turns; however, all other properties as a standard (in particular time stability, temperature coefficient, excitation independence, frequency
response) are degraded. Metrology-grade inductance standards have a non-
ferromagnetic core of high-dimensional stability; the maximum achievable Q
of self-inductors is limited (usually ≈ 10 at $1\,kHz$).

8.2.1 Self-inductance standards

A cylindrical winding, see Fig. 8.14, gives the highest Q. The resulting magnetic flux extends into space, and any ferromagnetic, or simply conducting
(because of induced eddy currents) object in the surroundings alters its value.
The winding cannot therefore be screened, and any magnetic flux in the environment (e.g., that caused by the measurement circuit itself, or by mains
currents) is linked to the standard and generates interference.

A measurement circuit suitable to handle cylindrical inductors usually has
an extended arm, sometimes meters long, provided with twisted or coaxial
connections; the standard is contacted at one side of the arm, and the measurement circuit lies at the other side. An inductance-inductance comparison
bridge has two arms, placed at 90° to minimize mutual inductance between
the standards being compared (Arri and Noce, 1974).

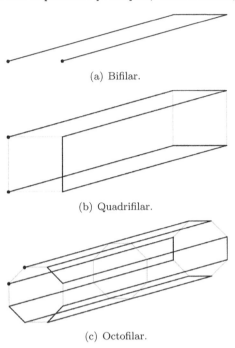

(a) Bifilar.

(b) Quadrifilar.

(c) Octofilar.

Figure 8.13
Wire arrangement of (a) bifilar, (b) quadrifilar, (c) octofilar calculable resistors.

A toroidal winding, see Fig. 8.15, gives a negligible flux outside the winding outer surface; it is therefore much less sensitive to external magnetic fields, and can be encased in a conducting shield. The quality factor is reduced with respect to the cylindrical winding.

Both cylindrical and toroidal inductors have a significant dependence of inductance over frequency, because of flux distortions caused by parasitic currents in the winding itself, and strong capacitive currents between winding turns. Resonant frequencies can be as low as a few kHz for high-valued (1 H range) inductors. Fig. 8.16 shows an example of the frequency dependence of a 1 H inductor up to resonance.

Most inductors are defined as 2T or 3T standards. Since the electromagnetic quantity inductance is defined only for a closed current path, an inductance bridge will measure the combination of the standard inductance and of the connections. For high- or medium-valued inductors, a *short* compensation (Sec. 4.8.2.3) can be applied, by short-circuiting the inductance connections. For inductors of very low value, since the short circuit can have and impedance comparable with the inductor itself, the inductance is defined as a difference value between readings from measurements involving additional binding posts (Fig. 8.17).

Figure 8.14
A Sullivan & Griffiths 400 mH four-terminal inductance standard, having a cylindrical winding. Year 1952.

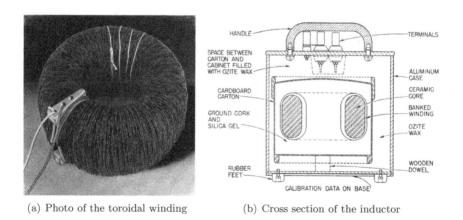

(a) Photo of the toroidal winding   (b) Cross section of the inductor

Figure 8.15
General Radio mod. 1482-A inductance standard. After Lamson (1952, Fig. 2 and 3), courtesy of the General Radio Historical Society.

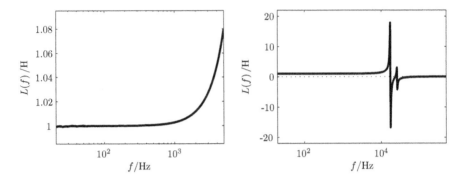

Figure 8.16
Inductance dependence over frequency, over two different frequency spans, for a 1 H toroidal, non-magnetic core inductor (General Radio mod. 1482-P). Measurements have been performed with an Agilent E4980A RLC bridge.

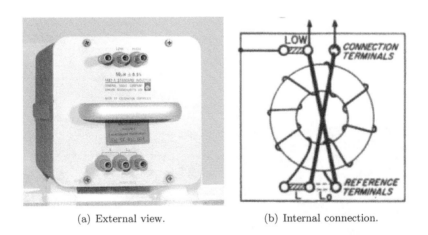

(a) External view.   (b) Internal connection.

Figure 8.17
(a) General Radio Mod. 1482-A 50 μH toroidal inductor, having six binding posts. (b) The internal connection of the inductor. The bridge is wired to the connection terminals as in three-terminal inductors (here the inductor is considered a two-terminal impedance, and the shield is connected to the LOW terminal). A first reading L is taken with the shorting link at the reference terminals in the L position (inductor connected). A second reading L_0 is taken with the shorting link in the L_0 position (so the inductor is disconnected). The inductor value is $L - L_0$. After Hersh (1960, Fig. 2), courtesy of the General Radio Historical Society.

The temperature coefficient of air-core self-inductors is essentially related to the thermal expansion of the materials involved, although for high-valued inductors the dielectric dissipation might play a role (Fiebiger and Dröge, 1984).

8.2.2 Mutual inductance standards

Mutual inductance standards have construction techniques similar to self-inductance standards. The impedance is defined as a 4T standard, see Fig. 8.18, but at variance with other standards the voltage terminals (the secondary coil) are isolated from the current terminals (the primary coil). Therefore, a typical additional definition condition is to maintain the low voltage and low current terminals at the same potential (often this condition is achieved by direct connection of these terminals). When measuring mutual inductance standards with modern impedance meters, care is required to avoid the effect of meter loading of the secondary winding: a possible solution involves buffer amplifiers (Bohacek et al., 2005).

Figure 8.18
A Sullivan & Griffiths 100 μH mutual inductance standard, year 1953.

8.2.3 Variable inductors, decade boxes

Self- and mutual variable inductors can be realized by having windings (or winding sections), which can be moved with respect to the other; the mutual inductance between windings (or sections) is a function of the position. A typical construction is shown in Fig. 8.19: two windings, a stator and a rotor, are mounted concentrically. A precision mechanism permit rotor control.

Decade box inductors also exist, see Fig. 8.20, realized with ferromagnetic-core individual inductors. If decades are set in the zero position, a residual inductance is still present, given by the switch inner wiring.

8.3 Capacitance standards

Capacitance standards are realized with metallic electrodes separated by a dielectric. The number of electrodes in parallel, their surface area, and the permittivity and thickness of the dielectric employed, determine the capacitance value and the corresponding dissipation factor. Electrolytic capacitors are typically not employed as capacitance standards, because of the poor stability and high temperature coefficient; high-value capacitance standards are realized by electromagnetic circuitry, Sec. 8.4.1 or by electronics, Sec. 8.4.3.

8.3.1 Gas-dielectric capacitors

Gas-dielectric capacitors are realized for capacitance values from $1\,\mathrm{pF}$ to $1000\,\mathrm{pF}$, although $10\,\mathrm{nF}$ high-voltage capacitors are available and $100\,\mathrm{nF}$ capacitors have been constructed in the past. For low-voltage capacitance standards the dielectric is typically air or dry nitrogen; high-voltage capacitors can employ different gases, e.g., SF_6, at high pressure, to increase the dielectric strength.

The capacitance stability is directly related to:

- the dimensional stability of the construction. Highly stable insulators (e.g., ceramics) and low-creep metals must be employed. The capacitance thermal coefficient is equal to the coefficient of linear thermal expansion α_L if the expansion is isotropic (Garcia-Valenzuela and Guadarrama-Santana, 2010). For typical metals, α_L is in the $10^{-5}\mathrm{K}^{-1}$ range, but can be lower for special alloys: it is $1.2 \times 10^{-6}\,\mathrm{K}^{-1}$ for invar. Capacitance is proportional to the electrode surface area but inversely proportional to electrode distance: therefore, alloys having different thermal expansion coefficients can be combined to achieve a non-isotropic thermal expansion and give a *compensated* geometry (McGregor et al., 1958).

- the stability of the permittivity of the gaseous dielectric, which is in turn related to temperature and pressure.[2] A construction within a sealed rigid vessel, such that in Fig. 8.21, permits to compensate for the pressure dependence; the use of a dry gas avoids also humidity dependence and pos-

---

[2]The capacitance C of a gas capacitor having a perfectly invariant geometry can be written as $C = \epsilon_0[1 + \chi_r(P,T)] \cdot K$, where ϵ_0 is the electric constant, $\chi_r(P,T)$ is the relative electric susceptibility of the gas at pressure P and temperature T, and K a characteristic geometrical length.

Since $\chi_r \ll 1$, the relative temperature and pressure coefficients of the variations of

Figure 8.19
General Radio mod. 107 variable inductor, inside view. Stator and rotor windings have separate terminals, so self-inductance can be achieved by connecting
them in series, anti-series, parallel, anti-parallel. After General Radio E catalog,
1928, p. 57, courtesy of the General Radio Historical Society.

Figure 8.20
4-decade inductance box, General Radio mod. 1490-D, 10 mH to 11.11 H with
1 mH resolution, configured as a 3T impedance. Each decade is composed of
four inductors (General Radio mod. 1481), wound on permalloy dust core,
having values $1, 2, 2, 5\times$ the decade step.

sible electrode corrosion. However, care must be taken to avoid the transmission of the inevitable strains of the vessel to the internal capacitor. The vessel is typically metallic and acts also as electrostatic screen.

Figure 8.21
General Radio mod. 1404-A 1000 pF gas-dielectric standard capacitor, with the external case removed. The cylindrical sealed vessel, filled with dry nitrogen, has been cut to show the capacitance plates inside. Courtesy of J. Schurr, PTB.

Low-value capacitors, 1 pF to 10 pF are often realized as cylindrical capacitors, see Fig. 8.22. Higher-value capacitors are realized by stacking parallel plates, Fig. 8.23. The typical construction is intrinsically three-terminal (al-

---

$C(P, T)$ with respect to some reference condition $C(P_0, T_0)$ can be written as

$$k_P = \frac{1}{C(P_0, T_0)} \left(\frac{\partial C(P, T)}{\partial T} \right) \approx \frac{\partial \chi_r(P, T)}{\partial T}; \qquad k_T = \frac{1}{C(P_0, T_0)} \left(\frac{\partial C(P, T)}{\partial P} \right) \approx \frac{\partial \chi_r(P, T)}{\partial P}.$$

For typical gases (air, nitrogen) near room temperature and pressure, one can assume that χ_r is proportional to its density: With the ideal gas law, we can write

$$\chi_r(P, T) = \frac{P}{P_0} \frac{T_0}{T} \chi_r(P_0, T_0),$$

which gives

$$k_T = -\frac{P}{T^2} \frac{T_0}{P_0} \chi_r(P_0, T_0); \qquad k_P = \frac{1}{T} \frac{T_0}{P_0} \chi_r(P_0, T_0).$$

For example, dry air at standard conditions ($P_{STP} = 0\,°C$, $T_{STP} = 100\,kPa$) has $\chi_{STP} \approx 5.9 \times 10^{-4}$, which gives at room temperature and pressure ($P = 23\,°C$, $T = 101\,kPa$) the coefficients $k_T = -1.9 \times 10^{-6}\,K^{-1}$ and $k_P = 5.4 \times 10^{-7}\,hPa^{-1}$.

though the definition given by external connection may be different) and include, in addition to the active electrodes, external (case), and internal shield electrodes: these collect fringe fields and can define the geometry. Care is taken to avoid that the electrostatic field between the active electrodes does not interact with any insulator employed in the construction.

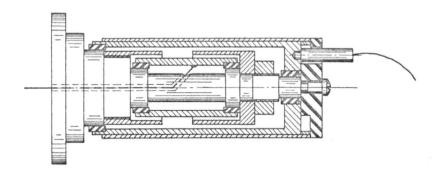

Figure 8.22
Cylindrical 1 pF capacitance standard. The capacitance length is defined by the screen between the active electrodes. ©2008 Canadian Science Publishing or its licensors. Reproduced, with permission, from Dunn (1964, Fig. 6).

Gas-dielectric capacitors have a small frequency dependence, since gases of small molecular weight have constant permittivity up to the GHz frequency range. The frequency dependence is dominated by the self-inductance: far from resonance it can be approximated by a quadratic dependence

$$C(f) = C_0 \left(1 + k\,f^2\right), \qquad (8.1)$$

where k can be either positive or negative; Fig. 8.24 gives an example of frequency dependence of an air-dielectric capacitor.

If the capacitor has a geometry which is highly symmetric , e.g., cylindrical coaxial (Awan and Kibble, 2007), a calculation of its frequency performance can be achieved with a distributed-parameter model. Otherwise, the constant k in Eq. (8.1) can be identified with resonance measurements, Sec. 4.9.

Related to gas-dielectric capacitors are *vacuum capacitors*, suitable for the construction of phase angle standards (Sec. 8.6) and, at cryogenic temperatures, as ultralow-loss dc capacitors in electron-counting capacitance standards (Sec. 9.4.2).

8.3.1.1 Low values

A particular construction of gas capacitors, useful to obtain very low-valued capacitance standards (down to the aF range), is given by the *Zickner ca-*

Figure 8.23
The electrodes of a General Radio mod. 1406-D 100 pF air-dielectric capacitance standard. Each electrode is composed of a number of circular plates supported by three conducting rods; the other electrode plates are interleaved with those of the first one, and supported by other three rods.

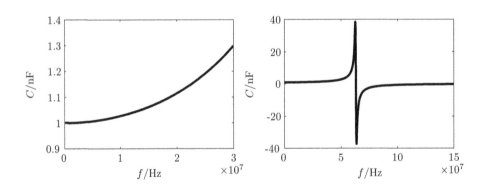

Figure 8.24
Frequency dependence of a gas-dielectric capacitor (Agilent mod. 16384A, 1 nF air-dielectric) for two different frequency ranges. Far from resonance (around 60 MHz) the frequency dependence is well approximated by Eq. (8.1).

pacitor (Zickner, 1930). A screen electrode is inserted in the gap between the active electrodes and intercepts the electric flux. A hole, or some holes in the screen electrode, allows a small portion of the flux between active electrodes to flow. Parallel-plate or cylindrical (see Fig. 8.25) geometry is typical; often, the screen electrode is removable and a set of plug-in electrodes, having holes of different sizes, are provided to obtain different capacitance values. The same

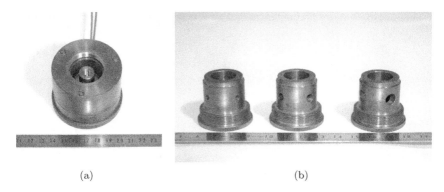

(a)   (b)

Figure 8.25
A Zickner capacitor, realized in cylindrical geometry. The capacitor (a) has two coaxial electrodes (the spike and the inner of the hole in the casing), screened by the body case. The electrical flux pass through the holes of the particular screen electrode, in the form of a plug (b), chosen from a set. The three plugs shown give a capacitance value of 20 aF, 1.6 fF, and 13 fF, respectively.

principle of electric flux interception can be realized with other geometries (Moon and Sparks, 1948).

8.3.2   Solid-dielectric capacitors

A solid dielectric capacitor can achieve a much higher capacitance than gas-dielectric ones, because

- the dielectric has a mechanical rigidity, so both the electrodes and the interelectrode spacing can be much smaller than in gas-dielectric capacitors. The minimum thickness is in practice limited by the dielectric strength, which typically ranges in the tens of $MV\,m^{-1}$.

- the relative permittivity of a solid dielectric is typically a few units (e.g., ≈ 5 for glass) instead of ≈ 1 of gases.[3]

---

[3]Although so-called *high-κ* dielectrics, such as *ferroelectrics*, can have a relative permittivity above 100, such materials are not employed in the construction of capacitance standards because of the poor stability over time and environmental conditions, and strong voltage dependence. For the same reason, *electrolytic* capacitors are not employed. The realization of very high-valued standard, above 1 mF, is achieved by synthesis (Sec. 8.4).

The time stability and the magnitude of environmental coefficients can be smaller than gas-dielectric capacitors. Frequency behavior is due to self-inductance (as in dielectric capacitors) but also to the frequency dependence of the permittivity of the dielectric, which often follows a "universal" power law decrease with frequency (Jonscher, 1977, 1992); the combination gives a typical non-monotonic behavior as shown in Fig. 8.26.

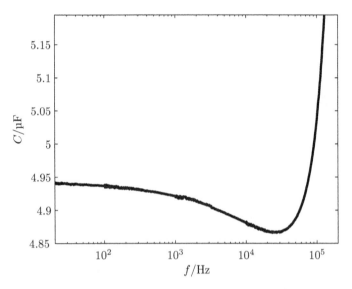

Figure 8.26
Frequency dependence of a solid-dielectric capacitor (Vishay mod. MKT-1822, metallized polyester film capacitor, $4.7\,\mu\text{F}$ nominal valued, measured with an Agilent E4980A RLC bridge).

Some solid dielectrics suitable for the realization of standard capacitors are:

fused silica or fused quartz is an extremely stable material, both electrically and mechanically, and is employed for the construction of low-value ($1\,\text{pF}$-$100\,\text{pF}$) standards. A design due to Cutkosky and Lee (1965) and followed in modern commercial realizations (Andeen-Hägerling mod. AH11A). Active and shield silver film electrodes are directly fired over a dielectric "hockey puck," Fig. 8.27; the element is then contacted through springs in a supporting cell, Fig. 8.28.

Both frequency dependence (Fig. 8.29) and loss ($<3 \times 10^{-6}$) at 1 kHz of fused silica capacitors are very small in the audio frequency range. Resulting temperature coefficient is around $12 \times 10^{-6}\,\text{K}^{-1}$ (Daniel, 1995). The long-term relative stability is typically better than 1×10^{-6} over one year.

Figure 8.27
Fused silica 10 pF capacitance element. The active electrodes are on top and
bottom faces of the dielectric cylinder, and a shield electrode is on the lateral
surface. The electrodes are given by a single silver film with isolating gaps
obtained by mechanically beveling the cylinder. The small indentation in front
is an adjustment to the nominal value. After Cutkosky and Lee (1965, Fig. 5). Official
contribution of NIST; not subject to copyright in the US.

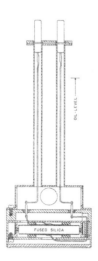

Figure 8.28
Cross section of a 10 pF fused silica capacitance standard, suitable for immer-
sion in an oil thermostat, showing the spring connection to the capacitance
element of Fig. 8.27. After Cutkosky and Lee (1965, Fig.1). Official contribution of
NIST; not subject to copyright in the US.

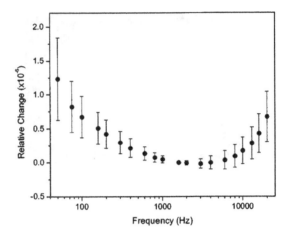

Figure 8.29
Frequency deviation of the capacitance of 10 pF fused silica standard, relative to its value at 1592 Hz, measured by comparison with a cross capacitor.
Reprinted with permission from Wang (2003, Fig. 3). ©2003, American Institute of Physics.

mica silicate minerals that can be cleaved in thin sheets, having high permittivity (≈ 7) and dielectric strength ($20\,\mathrm{MV\,m^{-1}}$ to $70\,\mathrm{MV\,m^{-1}}$), and a dissipation factor in the 10^{-4} range at audio frequency. Thin film metal (often silver) electrodes are deposited on mica sheets, which can be stacked and connected in parallel to increase the capacitance; the resulting pack is covered with resins (epoxy) and encased in a shield. Mica standard capacitors typically cover the range 1 nF to 1 µF.

plastic A wide range of plastics can be employed as dielectric material in capacitors. Plastics employed in standard capacitors are typically polystyrene, polycarbonate, teflon.

ceramics Ceramic chip capacitors like C0G/NP0 can have a temperature coefficient lower than $30 \times 10^{-6}\,\mathrm{K^{-1}}$ and a dissipation factor in the 10^{-5} range at kHz frequency, and can be employed to construct standards in the nF to µF range (Callegaro et al., 2005).

8.3.2.1 Variable capacitors and decade capacitance boxes

Mechanically variable capacitors are controlled by varying the amount of active surface area of the electrodes or the distance between them. A typical construction, Fig. 8.30(a), arranges semicircular metal plates on a rotor, interleaved (with appropriate air gaps) with plates on a stator; changing the rotor angle varies the capacitance between a minimum and a maximum. A reduction gear may be present to increase resolution.

Decade capacitance boxes, Fig. 8.30(b), have a construction very similar to decade resistance boxes; the last decade can be realized with a variable capacitor to increase the box resolution. The capacitance box is typically defined as a 3T standard with binding posts. When all decades are set to 0, the capacitance box still has a minimal capacitance value of a few pF given by all internal stray capacitances between contacts and connections. High-accuracy, remotely-controlled decade capacitance box suitable for primary metrology have been described (Eckardt et al., 1999; Wang and Lee, 2004).

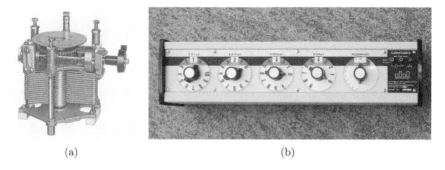

(a)   (b)

Figure 8.30
(a) Inner view of the General Radio mod. 222-L, 50 pF to 1500 pF variable air capacitor. Courtesy of the General Radio Historical Society. (b) 5-decade capacitance box JJ Lloyd mod. C500, up to 1.1111 μF. The last decade is a 100 pF continuously variable capacitor.

8.4 Synthesized impedance standards

Since the impedance of a standard is simply the transfer function between the input current and the output voltage at the appropriate terminals, any way to realize such transfer function, if satisfying the construction goals listed in Sec. 8.1.1 can be employed in the construction of an impedance standard. Here are some examples from commercial realizations.

8.4.1 Electromagnetic capacitors

High-value capacitance standards can be realized with the help of electromagnetic components. Fig. 8.31 shows an arrangement with two autotransformers having ratios r_V and r_I and a solid-dielectric capacitor of capacitance C_0.

With ideal autotransformers, the resulting four-terminal capacitor would have an apparent capacitance $C = (r_V r_I)^{-1} C_0$. Other arrangements are possible (Kuperman et al., 2010).

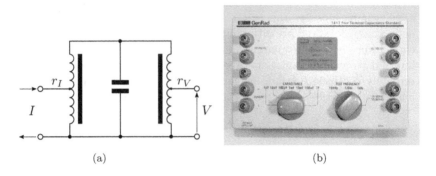

(a)　　　　　　　　　　　　　　　　　　　(b)

Figure 8.31
(a) An electromagnetic capacitance standard composed of two autotransformers and a capacitor. (b) The General Radio mod. 1417, $1\,\mu\text{F}$ to $1\,\text{F}$ electromagnetic capacitance standard.

8.4.2 Simulated inductors

Inductors have less ideal properties than resistance or capacitance standards, and in addition are bulky and expensive to construct. Electrical active networks (Riordan, 1967; Kalinowski, 1968; Johnson and Small, 1998; Horsky and Horska, 2002) can realize simulated inductance standards with properties competitive with real inductors, see, e.g., Fig. 8.32. They are often based on the implementation of a *gyrator* (Sec. 1.6.4.3) with operational amplifiers.

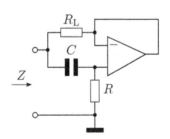

Figure 8.32
Simulated 2T inductor with an active op-amp and a RC differentiator. If $R_L \ll R$, the equivalent series inductance is $L = R_L RC$, and the series resistance is R_L.

8.4.3 Capacitance calibrators

Commercial multi-function calibrators, in addition to the more common functions (dc and ac voltage and current, resistance), can implement electronically a 2T or 4T capacitance function. An example is Fluke mod. 5500A, which synthesized capacitance can cover continuously a wide range, from pF to mF; the accuracy is in the 10^{-4} range but with strong limitations in the frequency range. The capacitance function is provided by a fixed capacitance standard and a gain amplifier, which gain can be digitally controlled, see Fig. 8.33.

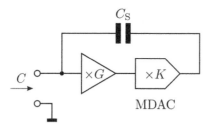

Figure 8.33
The synthesized capacitance function of Fluke mod. 5500A calibrator. Adapted from Fluke 5500, Fig 2.3. Capacitance $C = KGC_S$ can be varied continuously by adjusting the multiplying DAC gain K.

8.4.4 Impedance digital synthesis

The impedance standard can be simulated, by sampling the voltage (or current) excitation signal provided by the impedance meter and a real-time direct digital synthesis of the current (or voltage) corresponding signal of the desired simulated impedance value. Fixed impedance standards provide current/voltage conversion factors (Oldham and Booker, 1994).

8.5 Cryogenic standards

At audio frequency, normal air-core inductors have low-quality factors ($Q \approx 10$ at 1 kHz, mainly due to the dc resistance of the winding). Low-loss superconducting inductors, either cylindrical or toroidal, have been made, permitting the realization of tuned LC filters having Q exceeding 10^6 (Ricketts, 1976, 1978; Bonaldi et al., 1998; Pogliano et al., 2007).

Cryogenic vacuum-gap capacitors are employed in *electron-counting capacitance standard* experiments, see Sec. 9.4.2. Both parallel-plate (Zimmerman,

1996) and cylindrical (Overney et al., 2000) constructions have been experimented. The working temperature is in the 10 mK range. The dc leakage resistance is above $10^{19}\,\Omega$ (Zimmerman, 1996) and the expected frequency dependence below $2 \times 10^{-7}\,\mathrm{kHz}^{-1}$ (Zimmerman et al., 2006).

8.6   Standards of phase angle, time constant, etc.

Phase angle standards, *dissipation factor* standards, *time constant* standards, and *Q* standards are employed in the calibration of vector impedance meters and in electrical power metrology; the use of one name or another is a matter of convenience, since the same physical component can be considered a standard for all four quantities.

resistors The time constant τ of a resistor is the result of two contributions, a positive one given by its inductance and a negative one due to the distributed capacitance. In general, low-valued resistors (below $10\,\Omega$) have a positive phase angle, whereas high-valued ones (above $100\,\mathrm{k}\Omega$) give a negative phase angle. Rayner (1958) has shown that small resistor components in the range $10\,\Omega$dash$1000\,\Omega$ have a time constant below $1 \times 10^{-9}\,\mathrm{s}$.

calculable resistors The very same calculations that permit to obtain the resistance dependence over frequency of calculable resistors give also an estimate of their time constant. For dedicated standards, the geometrical dimensions can be chosen in order to minimize the time constant (Fujiki et al., 2003).

self inductors Losses in conductors and the core limit the quality factor. Special 2T self-inductors (e.g., Boonton Radio Corp. mod. 513-A) having a very stable Q value were employed in the past as reference standards for the Q-*meter* (Sec. 4.9).

mutual inductors The deviation of the phase angle from ideal $\pi/2$ value is caused by eddy current losses in the windings or the supporting structure, and the effects of intra- and interwinding capacitances and the associated losses (Butterworth, 1920). A suitable construction permit to keep losses in the 10^{-6} range up to kHz frequency (Hartshorn, 1925).

gas-dielectric and vacuum capacitors Loss factor at audio frequency of gas-dielectric capacitors is typically below 10^{-5} at 1 kHz. Several effects can be invoked to explain the residual loss, including

(a) plates and connection series resistance;

(b) dielectric losses in supporting insulators. A careful design such as that in Fig. 8.21 permits to eliminate such effect;

(c) resistive losses in the metal electrodes, which can increase at high frequency because of skin effect and eddy currents;

(d) mechanical losses due to vibrations induced by the electrostatic force;

(e) electromagnetic radiation;

(f) losses in dielectric films adsorpted or chemisorped on electrode surfaces.

Contrary to expectations, gas humidity does not seem to have direct effects on capacitor losses (Kouwenhoven and Lemmon, 1930). At audio frequency, it is believed that, in well-constructed gas-dielectric capacitors at low-operating voltage, the main cause of loss is (f). Special parallel-plate capacitors, which electrode spacing can be varied (Kouwenhoven, 1938; Astin, 1939; Inglis, 1975; So and Shields, 1979; Eklund, 1996) permit to perform absolute loss measurements, and therefore give loss angle standards having accuracies below 1×10^{-6}.

cross capacitor It has been shown theoretically (Shields, 1972) that the effects of axially uniform lossy dielectric films on any electrode of a cylindrical cross capacitor (Sec. 9.2.1.2) is equal and opposite for the two cross capacitance measurements. On a toroidal cross capacitor (see, e.g., Makow and Campbell, 1972), the compensation is less perfect, but the mean loss angle is below 1×10^{-8}.

compensated networks Secondary 3T, 2P, or 4P standards approaching 0 or $\pm\pi/2$ phase angle can be synthesized with passive networks. For example, the three-resistance network of Fig. 8.6 can be generalized to a RC network. Fig. 8.34 combines two capacitive resistors $Z_R = (R/2)(1 + j\omega\tau)$, where $\tau < 0$, and a capacitor C; by trimming $C \approx 4|\tau|/R$, a 2P/3T resistor having near zero time constant can be achieved. In a similar way, with two capacitors and a shunt resistors, a zero-loss capacitor can be achieved. In real networks, the compensation is achieved only at one specific working frequency, or in a restricted frequency range; the standard can be trimmed by comparison with a primary phase angle standard.

8.7 Open and short standards

Calibration of impedance meters, Sec. 4.8, typically requires *open* Z_O ($Z_O = \infty, Y_O = 0$) and *short* Z_S ($Z_O = 0, Y_O = \infty$) impedance standards. The construction of the standards depends on the impedance definition being applied for the measurements.

1P standards are typical of high-frequency measurements, Sec. 8.8. A

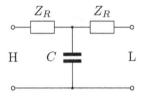

Figure 8.34
A *RC* network, constructed with resistors having negative time constant and a shunt capacitor, synthesizing a 3T/2P resistance standard having zero time constant.

scheme of 2P open and short standards is shown in Fig. 8.35. The physical realization of a 4P open and short standard is shown in Fig. 8.36.

(a) 2P open.　　　　　(b) 2P short.

Figure 8.35
Two terminal-pair open (a) and short (b) standards.

8.8 High-frequency and microwave standards

Primary impedance standards for high- and microwave frequencies are discussed in Sec. 9.2.2. Working impedance standards are 1P or 2P passive devices, grouped in *calibration* or *verification* kits and employed for routine calibration of automatic network analyzers (Sec. 4.11.1). Calibration kits typically include (Fig. 8.37, 8.38):

opens, 1P standards approximating the admittance value $Y = 0$. See also Sec. 8.7.

shorts, 1P standards approximating the impedance value $Z = 0$.

loads, 1P standards approximating the *reference impedance* Z_0 (see Sec. 1.6.3.5). In most cases, $Z_0 = 50\,\Omega$).

air lines, 2P standards approximating an ideal transmission line (Sec. 1.6.4.4).

Figure 8.36
A box that can act as an approximate 4P open or short standard. Note the
different labeling of 4P connections for the two configurations.

sliding terminations, 1P standards made of an air line which electrical
length can be mechanically varied by acting on a slide, terminated with
a load. The sliding load is employed by performing several measurements
and different line lengths; the resulting measurement set can be math-
ematically or graphically treated (with the help of the Smith chart) to
simulate the measurement on a reference impedance.

The actual nominal impedance value of each standard is given by stating the
values of the coefficients of an appropriate electrical model (Agilent 1287).
The typical model for 1P standards is a lossy transmission line, terminated
by an ideal impedance (a frequency-dependent lossless capacitor for the open,
a frequency-dependent lossless inductor for the short, Z_0 for the load).

Electronic calibration kits are multiport devices, including solid-state electronic devices, which impedance matrix can be changed under remote control.
The electronic calibration kit is controlled by the instrument, which execute
the steps of a calibration procedure. At the expense of a reduced accuracy,
the electronic calibration kit strongly reduces the number of connector mating
and unmating operations required for the calibration.

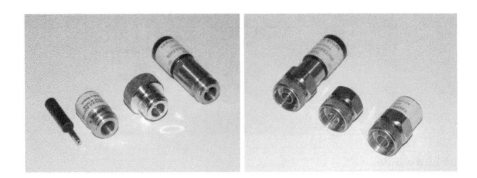

Figure 8.37
Open, short and $50\,\Omega$ load standards suitable for the SOLT calibration of
a network analyzer. (left) Female standards: Agilent 85032-60012 open (with
the associated isolated termination pin), 85032-60009 short, 00909-60010 load.
(right) Male standards: Agilent 85032-60007 open, 85032-60008 short, 00909-
60009 load.

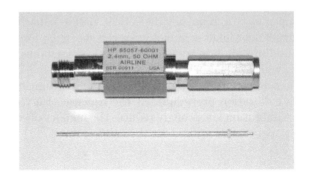

Figure 8.38
A $50\,\Omega$ coaxial air line, 50 mm nominal length, terminated with 2.4 mm connectors: Agilent 85057B-60001 is part of a verification kit for network analyzers
in the 45 MHz to 50 GHz frequency range.

9

Metrology: realization and reproduction

CONTENTS

9.1   The International System of Units

The International System of Units (in French: *Système International d'unités*, SI) is the modern form of the metric system, and is now globally adopted for the measurement of electromagnetic quantities.

The 11th CGPM laid down rules for the SI such as the prefixes, the derived units, etc. The SI is not static but evolves to match the world's increasingly demanding requirements for measurement.

The brochure entitled *The International System of Units* is the essential reference for all those who wish to use the SI correctly. It contains the official definitions of the base units of the SI, and all the decisions of the CIPM and CGPM related to the SI, its formalism and use. The brochure is periodically updated. The most recent edition is the 8th (SI brochure).

Other systems of units (most notably CGS, the *centimetre-gram-second* system and its electrostatic, electromagnetic, and Gaussian variants) are now considered obsolete.

9.1.1 Base and derived units

In the SI, electromagnetic units are the base unit ampere and several derived units; these are defined as product of powers of the four base units meter (m), kilogram (kg), second (s), and ampere (A); that is, all have the form $m^\alpha \, kg^\beta \, s^\gamma \, A^\delta$, where α, β, γ, and δ are integers. Just for convenience, certain derived units have been given special names and symbols: those related to electromagnetism are listed in Appendix A, Tab. A.1. These special names and symbols may themselves be used in combination with the names and symbols for base units and for other derived units to express the units of other derived quantities: for example, permittivity is expressed in Fm^{-1}.

9.1.2 Prefixes

The SI adopts a series of prefix names and prefix symbols to form the names and symbols of the decimal multiples and submultiples of units, ranging from 1×10^{24} to 1×10^{-24}; they are listed in Appendix A, Tab. A.2. The expression of the value of electromagnetic quantities benefits large or small prefixes, more often than in other scientific fields. For example, it is common to speak of fA current, $P\Omega$ resistance, or aF capacitance values.

9.1.3 Definition of units

In SI, the definition of the base unit ampere is mechanical:

> The ampere is that constant current which, if maintained in two straight parallel conductors of infinite length, of negligible circular cross-section, and placed 1 metre apart in vacuum, would produce between these conductors a force equal to 2×10^{-7} newton per metre of length. (SI brochure, 2.1.1.4.)

Since the definition of ampere is mechanical in its essence, it follows that electromagnetic derived units have an ultimately mechanical definition also.

Further, from the definition of the ampere and that of the meter (in turn given in terms of the second and of the speed of light in vacuum, the exact quantity $c = 299\,792\,458\,\mathrm{m\,s^{-1}}$), it follows that:

- the magnetic constant, μ_0, also known as the permeability of free space, is an exact quantity: $\mu_0 = 4 \times 10^{-7}\,\mathrm{H\,m^{-1}}$;

- the electric constant, ϵ_0, also known as the permittivity of free space, is an exact quantity: $\epsilon_0 = \left(\mu_0 c^2\right)^{-1} = 8.854\,187\,817\ldots\,\mathrm{F\,m^{-1}}$;

- the impedance of free space, Z_0, is an exact quantity: $Z_0 = \mu_0\,c = \sqrt{\mu_0\,\epsilon_0^{-1}} = 376.730\,313\,4\ldots\,\Omega$.

The fact that μ_0 and ϵ_0 are exact quantities is the basis for the realization of SI units of impedance, see Sec. 9.2.

9.2 Realization of impedance units

The practical *realization* of a SI unit definition is bound to a physical experiment and has therefore an uncertainty larger than zero. The realization of impedance units ohm, henry, and farad is based on the SI exact value of the magnetic and electric constants μ_0 and ϵ_0.

9.2.1 Low frequency

The (self- or mutual) inductances and capacitances of a system of electrical conductors, in absence of dielectric or magnetic materials, can be calculated from geometrical properties alone.

If the geometry of the system of conductors is sufficiently simple, explicit mathematical expressions for their inductance or capacitance value may exist. To give two elementary examples:

- the low-frequency inductance L of a circular conductive loop (of radius r), made of a circular perfect conductor (of radius a), in vacuum, is $L = \mu_0\,r\,[\log(8r/a) - 7/4]$;

- the capacitance C of a conducting sphere of radius R in vacuum is $C = 4\pi\epsilon_0 R$.

The previous examples are not adequate for a practical impedance realization, which require a careful choice of the calculable geometrical shape of conductors in order to minimize:

- the dependence of L or C on inevitable deviations of the mechanical realization of conductors' shapes from the ideal geometry employed in the mathematical modeling;

- the number, and practical difficulty, of the accurate length measurements, which are needed in the calculation.

9.2.1.1 Calculable inductor

Calculable self- or mutual inductors have been used until 1970s for the realization of the ohm in several national metrology institutes.

The most accurate realizations have been performed at PTB (Linckh and Brasack, 1968) with a 100 mH cylindrical self-inductor (Fig. 9.1), and at NPL (Harrison and Rayner, 1967) with a 10 mH Campbell mutual inductor (see below). Both realizations approach a relative uncertainty of 1×10^{-6}.

Figure 9.1
A photo of the PTB calculable self-inductor. 945 copper turns are woven in a single layer on a porcelain cylinder. After Linckh and Brasack (1968, Fig. 3), ©IOPP.

The Campbell mutual inductor (Campbell, 1907) is a particular configuration aiming at the minimization of the dependence of the mutual inductance on the geometry of the secondary winding, and on its imperfect coaxiality with the primary winding. Campbell's own realization is shown in Fig. 9.2 and 9.3.

9.2.1.2 Calculable capacitor

The discovery of a new theorem in electrostatics (Thompson and Lampard, 1956; Lampard, 1957; Thompson, 1959; Lampard and Cutkosky, 1960) led to the realization of the *calculable cross-capacitor*. The theorem can be stated as follows (adapted from Lampard and Cutkosky (1960)):

> Consider a conducting cylindrical shell whose right cross section, shown in Fig. 9.4, is the arbitrary closed surface S. Let this shell be divided into four parts 1, 2, 3, 4 by infinitesimal gaps. The capacitance per unit length C_{13} between electrodes 1 and 3, and the capacitance C_{13} per

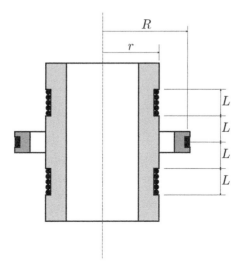

Figure 9.2
Campbell's mutual inductor diagram. The primary circuit consists of two equal single-layer coils put in series, of radius r and height L, wound on the same cylinder with a gap of height $2L$ between them. The multilayer secondary winding is coaxial with the primary coils and placed in the middle of the gap; its radius $R = 1.46r$ is chosen to maximize the mutual inductance. Adapted from Campbell and Childs (1935, Fig. 140, p. 149).

unit length between electrodes 2 and 4 are related by the expression[1]

$$\exp\left(-\pi\epsilon_0 C_{13}\right) + \exp\left(-\pi\epsilon_0 C_{24}\right) = 1. \tag{9.1}$$

If there is sufficient symmetry such that $C_{13} = C_{24} = C$, from Eq. (9.1) follows that

$$C = \epsilon_0 \frac{\log 2}{\pi} = 1.953549043\ldots \times 10^{-12}\,\mathrm{F\,m^{-1}} \qquad \text{[exact]}. \tag{9.2}$$

A didactical proof of the theorem is given by Jackson (1999).

It has been shown (Lampard and Cutkosky, 1960) that the effect of dielectric layers on electrode surfaces, in conditions of symmetry, is negligible. The effect of deviations from the perfect cylindrical symmetry of the construction has been analyzed by several papers (Pogliano, 1990; Fiander and Small, 2008).

---

[1] A magnetostatic dual of Eq. (9.1) can be written (Frenkel, 1993) for the same geometry, where equipotential perfect conductors (superconductors) 1, 2, 3, 4 are crossed by axial surface currents. However, no standard has yet been constructed on such principle.

Figure 9.3
A photo of Campbell's mutual inductor at the National Physical Laboratory.
After Campbell and Childs (1935, Fig. 141, p. 149).

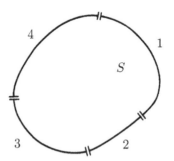

Figure 9.4
The general geometry of four conductors 1, 2, 3, 4 having cylindrical symme-
try, and arranged in a closed shell with infinitesimal gaps, analyzed by the
Thompson-Lampard theorem.

After the discovery of the theorem, several metrology institutes engaged in the construction of cross-capacitors.

For practical reasons, in the calculable capacitors employed in the farad realization the four electrodes are defined by circular cylinders of equal diameter. The first realizations were based on electrodes having a fixed length, with guards at their end in order to reduce the effect of fringe fields (McGregor et al., 1958; Cutkosky, 1961; Dunn, 1964). An example of such capacitor is shown in Fig. 9.5.

Figure 9.5
Fixed calculable capacitor, constructed from stacked gauge bars, developed at NRC. ©2008 Canadian Science Publishing or its licensors. Reproduced, with permission, from Dunn (1964, Fig. 13).

In order to improve the electrode length definition and its measurement, cross-capacitors of more recent realization employ cylindrical guard electrodes, which delimit the active length of the capacitor, as shown in Fig. 9.6. A state-of-the-art realization is shown in Fig. 9.7.

Near the guard electrodes the distribution of the electric field has no more translational symmetry, so the cross-capacitor theorem cannot be applied. However, one of the guard electrodes can be moved axially, so the active length of the cross-capacitor can be varied. Under the assumption that the distorted field geometry is simply axially translated together with the guard electrode movement, the cross-capacitor theorem can still be applied to the remaining active length. The capacitance difference between two guard positions can be therefore still computed from the distance between the positions from Eq. (9.1) or Eq. (9.2). The motion of the guard electrode can be measured with high accuracy using a laser interferometer.

The low capacitance value, typically less than 1 pF, given by practical calculable capacitors, asks for a dedicated bridge for its measurement and

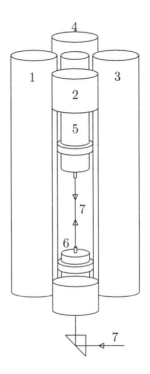

Figure 9.6
Cross capacitor with movable guard electrode. 1, 2, 3, and 4 are the four
cylindrical electrodes to which the cross-capacitor theorem applies; 5 and 6
are the two guard electrodes; electrode 6 can be moved axially between two
positions; the motion is monitored by a laser interferometer 7.

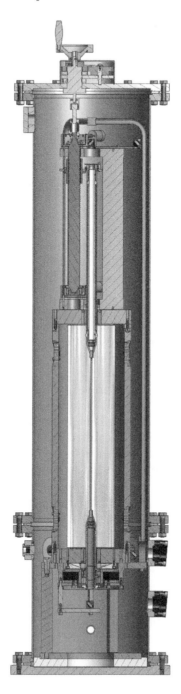

Figure 9.7
A longitudinal section of the NMIA-BIPM cross capacitor. The section shows
the four cylindrical main electrodes and the handwheel-driven guard electrode
within the vacuum enclosure, which has viewports for the interferometer beam
(not shown). Courtesy of J. Fiander, NMIA.

scaling to maintained capacitance standard, typically at 10 pF or 100 pF level (Kibble and Rayner, 1984, Fig. 6.18).

The recent development of nanofabricated devices, and the corresponding very small capacitances involved, has raised interest in the realization of the capacitance unit in the fF range. NIST is developing a small-size calculable capacitor having values in the 100 aF–10 fF range, with a target uncertainty of 100 yF (Durand et al., 2010).

9.2.2   High frequency

Impedance traceability in the radio and microwave frequency range is given by the mechanical measurement of all relevant dimensions of passive components constructed with high-conductivity and mechanically stable metals (typically gold-plated copper alloys are employed) and low-loss dielectric (typically air). The mechanical measurements can be performed by air gauging (through measurement of gas flow rate (Ide, 1992)), mechanical or laser gauges, and/or with three-dimensional coordinate-measuring machines.

The length measure set, and assumption on electromagnetic properties of the materials involved (conductivity, air dielectric constant and permeability) are the input of a model (analytically solvable, or computed with numerical methods). The model output is a set of impedance-related quantities such as the characteristic impedance, the scattering parameter matrix and so on.

Unsupported, or *beadless*, air-dielectric coaxial lines are the most precise reference standard for RF & MW impedance. A typical construction, see Fig. 9.8, involves two metallic parts, an outer (a straight cylinder terminating at each end with the outer part of the chosen connector type) and an inner (a straight circular rod). During the construction of the measurement circuit, the inner part is inserted in the outer and is coaxially supported at each end by the inners of the connectors belonging to the external circuitry.

The simplest model for a coaxial air line assumes perfect conductivity and perfect cylindrical symmetry. If d is the external diameter of the inner conductor, D is the internal diameter of the hollow external conductor, ℓ is the geometrical length, ϵ and μ are the relative dielectric constant and permittivity of air,[2] the characteristic impedance of such ideal line is (Sec. 3.4.1)

$$Z_{\mathrm{L}} = \frac{1}{2\pi}\sqrt{\frac{\mu}{\epsilon}}\log\left(\frac{D}{d}\right),$$

and the corresponding impedance matrix at angular frequency ω is (Sec. 1.6.4.4)

$$Z_{11} = Z_{22} = Z_{\mathrm{L}}\coth(\gamma\ell); \qquad Z_{12} = Z_{21} = Z_{\mathrm{L}}\operatorname{csch}(\gamma\ell),$$

---

[2]For air at room pressure and temperature, ϵ is larger than ϵ_0 by less than one part in 10^3; expressions for calculation of ϵ in terms of its temperature, pressure, and humidity have been published (Essen and Froome, 1951; Thayer, 1974). μ can be considered equal to μ_0 within one part in 10^6 (Essen and Froome, 1951).

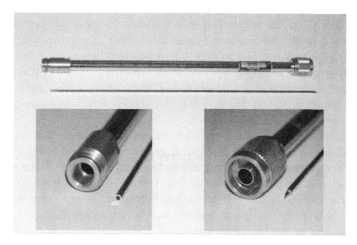

Figure 9.8
Unsupported coaxial line, N connector, 200 mm length. Rosenberger mod. RPC-N, custom.

where $\gamma = j\omega\sqrt{\epsilon\mu}$ is its *propagation constant*. Similar expressions can be written for the admittance or scattering parameter matrices.

More refined explicit expressions can take into account the effect of finite conductivity (Harris and Spinney, 1964) and diameter inhomogeneities along the line (Kossel et al., 2004; Leuchtmann and Rufenacht, 2004). Deviations from perfect cilindricity and surface roughness have also been considered (Leuchtmann and Rufenacht, 2004; Holt, 2009).

Other standards which can be modeled in a similar way, although with larger uncertainties, are one-port *opens* and *shorts* (Sec. 8.8) (Bianco et al., 1980).

Air-dielectric coaxial lines are limited in the maximum available length (typically <300 mm), which in turn limits their usefulness as impedance standard to frequencies above tens of MHz.

9.3   Reproduction of the ohm

9.3.1   The quantum Hall effect

The *quantum Hall effect* was discovered by von Klitzing et al. (1980). It is a quantum-mechanical version of Hall effect, observed in *two-dimensional electron gas* (2DEG, where electrons are free to move in two dimensions but tightly confined in the third), at very low temperature and high magnetic fields.

The 2DEG can be created in semiconductor devices by confinement with electrostatic fields. Silicon metal-oxide field-effect transistors (MOSFET) require an appropriate gate voltage polarization, whereas in gallium arsenide heterostructures the field is created at the interface of two epitaxial layers of different composition. Another physical system where 2DEG naturally occurs is in *graphene*, one-atom thick planar structures of carbon atoms arranged in a two-dimensional hexagonal lattice.[3]

For metrological applications, the active part of the device is constructed in the form of a bar, see Fig. 9.9, with contacts at its end (source and drain) where the measurement current I is injected, and a number of contacts on each long side (usually three per side). The voltage drop V_{xx} measured between two contacts on the same side of the bar defines the magnetoresistance $R_{xx} = V_{xx}/I$. The voltage V_{xy} measured across two corresponding contacts on the opposite sides of the bar defines the Hall resistance $R_H = R_{xy} = V_{xy}/I$.

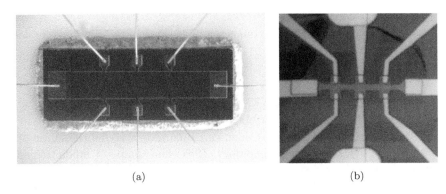

(a)   (b)

Figure 9.9
(a) Optical micrography of a AlGaAs/GaAs Hall bar heterostructure (Laboratoire Electronique Philips LEP 514). The bar has a length of 1 mm and a width of 0.4 mm; two current and six voltage AuGeNi/TiPtAu contacts are bonded with gold wires to the sample holder (not visible). Courtesy of C. Cassiago, INRIM. (b) Optical micrography of a graphene Hall bar, having a length of ≈15 μm and a width of ≈2 μm. Courtesy of S. Borini, INRIM.

The theory of classical Hall effect gives for a 2DEG having a carrier surface concentration n_s under a magnetic field density B, a linear relation between R_H and B

$$R_H = \frac{B}{n_s\, e}, \tag{9.3}$$

where e is the elementary charge.

---

[3]Quantum Hall effect in graphene has been observed even at room temperature (Novoselov et al., 2007); at low temperature, the feasibility of ohm reproduction with graphene samples has been demonstrated (Janssen et al., 2011).

The quantum Hall effect is shown in Fig. 9.10: the perturbation of such linear dependence with the appearance of flat plateaux in R_H. For the same field, the magnetoresistance R_{xx} goes to zero.

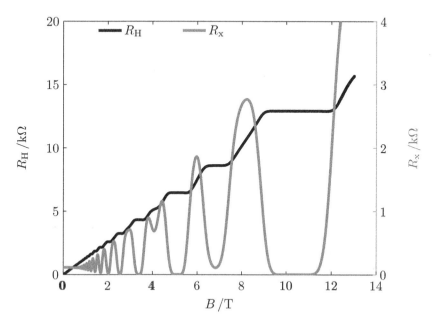

Figure 9.10
Observation of quantum Hall effect on the sample of Fig. 9.9(a). The sample is at a temperature of 2.53 K and carries a dc current of 25 μA. Voltage V_H across two opposite contacts transverse to the current flow, and voltage V_x between two adjacent contacts, is measured as a function of the magnetic field B. The graph shows the value of the Hall resistance $R_H = V_H/I$ and of the longitudinal resistance $R_x = V_x/I$. Hall plateaus corresponding to $R_H = R_K/i$ (with $i = 2, 4, 6...$) are clearly visible; correspondingly, R_x goes to zero. Data courtesy of C. Cassiago, INRIM.

According to theory (von Klitzing, 1986) each plateau, identified by a positive integer i, is centered on magnetic field values $B_i = n_s h/i\,e$, and has a resistance value R_K/i, where R_K is the *von Klitzing constant*:

$$R_K = \frac{h}{e^2} = \frac{\mu_0\,c}{2\alpha}. \tag{9.4}$$

R_K is thus linked to the fine structure constant α, which can be measured by non-electrical means (CODATA 2010).

9.3.2 R_K value(s)

The CODATA 2010 least-squares adjustment, which embeds Eq. (9.4) as an exact physical law, gives the value (App. C)

$$R_K = 25\,812.807\,443\,4(84)\,\Omega \quad [3.2 \times 10^{-10}].$$

CODATA 2010 conducted an investigation of the exactness of Eq. (9.4), by performing another adjustment where a weaker relation

$$R_K = \frac{h}{e^2}(1 + \epsilon_K)$$

is assumed, and ϵ_K is an unknown correction factor, which results from this new adjustment. Its 2010 value is $\epsilon_K = 26(18) \times 10^{-9}$.

In order to take advantage of quantum Hall effect to maintain laboratory reference standards of the electrical units, and at the same time taking care not to change their SI definitions, the 18th CGPM in 1987 adopted Resolution 6 which calls for representations of the ohm to be based on conventional values for R_K. On January 1, 1990 (Taylor and Witt, 1989), the conventional value $R_{K-90} = 25\,812.807\,\Omega$ of the von Klitzing constant was adopted. To the choice of R_{K-90}, impedance conventional units Ω_{90}, H_{90}, and F_{90} are associated, having values[4]

$$\Omega_{90} = [1 + 1.718(32) \times 10^{-8}]\,\Omega,$$
$$H_{90} = [1 + 1.718(32) \times 10^{-8}]\,H,$$
$$F_{90} = [1 - 1.718(32) \times 10^{-8}]\,F.$$

9.3.3 Reproduction of the ohm in dc and ac regime

Measurements in the dc regime of the quantum Hall resistance are routinely performed by national metrology institutes. Typical measurements involve gallium arsenide heterostructure devices, which require cooling below 2 K and B in excess of 10 T to reach the $i = 2$ plateau.

After a number of standardized tests, to ensure good quantization conditions (Delahaye, 1989; Delahaye and Jeckelmann, 2003), a standard resistor can be calibrated in terms of the quantized Hall resistance (typically by employing currents below 50 μA). The calibration can be performed with a dc potentiometer, or with a current comparator (Witt, 1998; Jeckelmann and Jeanneret, 2003).

Of course, if the calibrated standard has a known frequency dependence, for example, if it is a calculable resistor (Sec. 8.1.2.6), it can be employed after calibration in ac measurement setups.

---

[4]The conventional units Ω_{90}, H_{90}, and F_{90} are written in italic type in recognition of the fact that they are physical quantities.

To improve the uncertainty in the reproduction of the ac resistance unit, efforts to measure the QHE in the ac regime have been attempted in the last 20 years. However, despite the great care in undertaking the construction of special ac bridges and probes, unexpected dependencies of the QHE plateau resistance versus the measurement frequency and current appeared. Only very recently (Overney et al., 2006; Schurr et al., 2007), these dependencies have been definitively ascribed to capacitive effects. The effects can be corrected by guard electrodes (Overney et al., 2006; Schurr et al., 2007; Kibble and Schurr, 2008), and a new guideline to perform accurate QHE measurements at audio frequency finally appeared (Ahlers et al., 2009).

It has been shown (Schurr et al., 2009) that it is possible to employ two independent quantum Hall devices in a quadrature bridge, and calibrate capacitance standards with unprecedented accuracy (6×10^{-9}), giving the perspective of new measurements of the von Klitzing constant.

9.4 Counting electrons

9.4.1 Single-electron tunneling devices

Single-electron tunneling is a quantum phenomenon that permits to control the number of electrons stored in an isolated conducting island, by individually moving one electron at a time in or out of the island via individual electron tunneling events. The physical basis of single-electron tunneling devices is the phenomenon of *Coulomb blockade*, described in Sec. 9.4.1.1. The Coulomb blockade principle can be employed in devices constructed with nanoscale island, tunnel junctions, and capacitively coupled electrodes, as those of Sec. 9.4.1.2 and 9.4.1.3.

Single-electron tunneling devices and their application in metrology is the subject of a number of review papers (see, e.g., Likharev, 1999; Keller, 2001; Piquemal et al., 2007). The construction of planar single-electron devices ask for advanced nanolitography techniques, see Amato and Enrico (2011) for a review.

9.4.1.1 Coulomb blockade in a single-electron box

Consider the *single electron box* of Fig. 9.11. A small metallic island is coupled through a tunnel junction and a capacitive gate to a voltage bias generator V. Via tunneling events, electrons charge the island with charge $Q_i = -n\,e$, where n is an integer and e the charge quantum. The gate has capacitance C_G and holds charge Q_G; the tunnel junction has tunnel resistance R_T, capacitance C, and holds charge Q; then, $Q_i = Q - Q_G$.

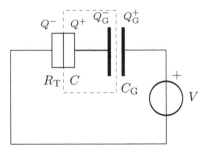

Figure 9.11
Single-electron box, coupled to an external circuit with a tunnel junction (with tunnel resistance R_T and capacitance C), and a capacitor C_G.

Circuit analysis of the mesh of Fig. 9.11 gives

$$Q = \frac{C}{C + C_G}(C_G V + Q_i),$$

$$Q_G = \frac{C_G}{C + C_G}(C V - Q_i),$$

from which the electrostatic energy E,

$$E = \frac{Q^2}{2C} + \frac{Q_G^2}{2C_G} = \frac{C C_G V^2 + Q_i^2}{2(C + C_G)},$$

the generator work $W = Q_G V$, and the free energy F

$$F = E - W = \frac{(C_G V + Q_i)^2}{C + C_G} + K = \frac{(C_G V - n e)^2}{C + C_G} + K \qquad (9.5)$$

can be computed (K is a constant term).

At equilibrium at a given bias V, the minimization of the free energy $F(V)$ given by Eq. (9.5) gives the corresponding equilibrium electron occupation of the box $n(V)$, which has the staircase appereance shown in Fig. 9.12.

In the derivation above, two hypotheses have been made:

1. The spacing between energy levels of the single electron box is large with respect to the average thermal excitation: $\frac{e^2}{2(C + C_G)} \gg k_B \Theta$. With nanofabrication techniques, device capacitances in the fF range can be achieved; an adequate working temperature lies in the tens of mK range.

2. The spacing between energy levels of the single electron box is large with respect to the energy uncertainty of an occupation state, in turn related by uncertainty principle to the state lifetime $R_T C$ caused by tunneling events. This gives the condition $R_T \gg R_K$, where R_K is the von Klitzing constant (Sec. 9.3.2).

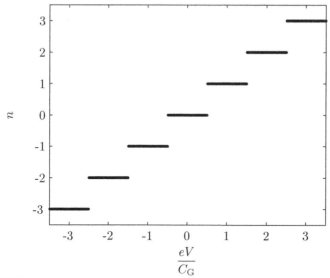

Figure 9.12
Single-electron box occupation number n versus applied bias voltage V.

9.4.1.2 Single-electron transistor

In a single-electron transistor (SET), Fig. 9.13 (Averin and Likharev, 1986; Fulton and Dolan, 1987) the island of the electron box is coupled with *two* tunnel junctions to external electrodes, called *source* and *drain*. The island is also capacitively coupled (with capacitance C_G) to a *gate* electrode. In operating conditions, the source-drain bias voltage V and gate voltage V_G can be set with voltage generators.

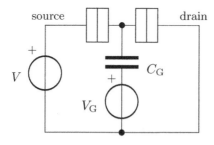

Figure 9.13
Single-electron transistor (SET).

In order to have a nonzero source-drain current I, each electron has to cross the two tunnel junctions: for example, if $V < 0$, source to island, and then island to drain. Both these crossings are regulated by the Coulomb blockade phenomenon, can occur or not depending on the corresponding free-energy difference between initial and final state, and can be analyzed with calculations similar to Sec. 9.4.1.1.

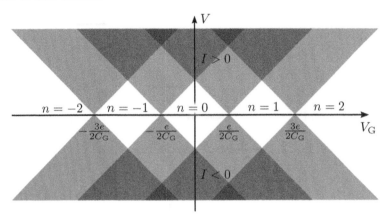

Figure 9.14
SET conductance versus source-drain voltage V and gate voltage V_G ("SET diamonds"). On white regions, conductance is zero and n is the number of electrons in the island. On shaded regions, $I \neq 0$.

The result of the calculations is shown in Fig. 9.14. For $V_G = 0$, current flow is inhibited except for particular bias voltage values, $V = (2k+1)\dfrac{e}{2C_G}$, $k = \pm1, \pm2, \ldots$, which are spaced by $\Delta V = \dfrac{e}{C_G}$, the "Coulomb gap" width. For $|V_G| > 0$, the set of V values that permits conduction widens from a point to an interval, and the Coulomb gap is progressively reduced, until it disappears. Hence, if properly polarized, the device can act as a transconductance amplifier, with a behavior similar to a field-effect transistor: a variation of V_G causes a variation of I.

A SET can sense a charge of $10^{-3}e$ on the gate electrode and is therefore an extremely sensitive electrometer.

9.4.1.3   Single-electron pump

The single-electron pump (Pothier et al., 1992) is a device, also based on Coulomb blockade phenomenon, which acts as a controlled current generator, by moving electrons from source to drain one at a time under an external electrical command. A three-junction single-electron pump is shown in Fig. 9.15. The structure of the SET is replicated: two islands in sequence, a and b, are coupled (between them, and to source and drain), with three tunnel junctions 1, 2, and 3. Each island has is own control gate electrode, biased to voltage $V_{G\,a,b}$.

Suppose the initial electron population number of islands a and b are n_a and n_b; this pump state can be written (n_a, n_b). Appropriate choices of the gate voltages (V_{Ga}, V_{Gb}) can allow the tunneling of *one* electron through a

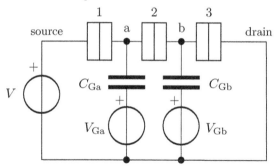

Figure 9.15
A three-junction single-electron pump.

junction (tunneling of more than one electron being inhibited by Coulomb blockade phenomenon), by at the same time keeping the other two junctions "shut" (that is, making a tunneling event energetically unfavorable on these junctions). It is possible to realize a pump cycle, i.e., the transfer of an electron from source to drain, with a sequence of three steps, each one realized with a corresponding choice of (V_{Ga}, V_{Gb}):

1. $(n_a, n_b) \rightarrow (n_a + 1, n_b)$, move an electron through junction 1 from source to a;

2. $(n_a + 1, n_b) \rightarrow (n_a, n_b + 1)$, move an eelctron through junction 2 from a to b;

3. $(n_a, n_b + 1) \rightarrow (n_a, n_b)$, move an electron through junction 3 from b to drain.

The typical working condition for the pump is $V = 0$; the work done on the electrons is performed by the gate voltage generators. The pump is reversible, simply by reversing the (V_{Ga}, V_{Gb}) sequence.

The charge moved after n pump cycles is an integer multiple $Q = -ne$ of the electron charge e. If pump cycles occur at the steady rate f, the corresponding pumped electric current is $I = ef$.

SET pump accuracy is limited by three phenomena:

thermal errors: electron thermal energy can induce uncontrolled tunneling events. Thermal errors can be reduced by lowering the experimental base temperature. However, electron heating events can occur because of the absorption of microwave photons from the electrical wiring, which must therefore heavily shielded and filtered.

frequency errors: the counting speed is limited by the tunnel rate, so $f \ll (R_T C)^{-1}$ (where R_T and C are the junctions' resistance and capacitance) for a proper pump working. With present technologies, $R \sim 100\,\mathrm{k\Omega}$ and $C \sim 1\,\mathrm{fF}$ give a limit of $f \sim 10\,\mathrm{MHz}$, corresponding to I in the pA range;

co-tunneling events, the simultaneous tunneling of electrons from source to drain through both junctions. The effect on the generated charge can be lowered by increasing the number of islands and junctions, as in the NIST 7-junction pump (Keller et al., 1996), or by on-chip resistors in series with the pump, the so-called R-pump (Zorin et al., 2000).

9.4.2   Electron-counting capacitance standard

In *electron-counting capacitance standard* (ECCS) (Keller et al., 1999), the definition of capacitance $Q = CV$ is directly employed in dc regime.

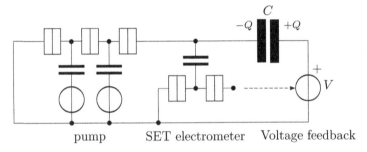

pump   SET electrometer   Voltage feedback

Figure 9.16
Electron-counting capacitance standards block schematics, using single-electron devices described in Sec. 9.4.1. The single-electron pump (Fig. 9.15) charges capacitor C to Q; a SET electrometer (Fig. 9.13) nulls low voltage side of C by driving a feedback generator to voltage V.

Looking at Fig. 9.16, a single-electron pump charges a capacitor C with n electrons to charge $Q = -n\,e$. A feedback generator, driven by a SET electrometer, permit the 3T definition of C and to voltage buildup on the pump. Voltage V on the capacitor is measured with a voltmeter, typically based on a Josephson DAC (Sec. 5.2.3). The capacitance is determined as

$$C = \frac{Q}{V} = \frac{-n\,e}{V}.$$

After the ECCS calibration phase, the capacitor can be electrically disconnected from the cryogenic devices with a system of needle switches (not shown in Fig. 9.16) and connected to a room-temperature capacitance bridge, working at audio frequency, in order to perform comparisons with room-temperature capacitance standards.

The accuracy of an ECCS experiment is limited, in addition to the properties of the single-electron devices involved, by the validity of the assumptions:

- the entire charge Q is stored in the capacitor for the duration of the experiment without any dissipation;

- the frequency dependence of C must be negligible, or known;

- C must be stable between ECCS and room-temperature bridge mesurement phases.

Experimental and modeling work give evidence that vacuum-gap cryogenic capacitors (see Sec. 8.5) achieve such extreme needs.

The present status of ECCS experimental relative uncertainty is in the order of a few parts in 10^7 (Keller et al., 2007). The ECCS experiment is not yet considered as a viable reproduction of the capacitance unit for traceability issues, but as one method to achieve the so-called *quantum metrology triangle*, see Sec. 9.4.3.

9.4.3 Quantum metrology triangle

The quantum metrology triangle (Keller, 2008, and refs. therein) is a consistency test of three quantum electrical standards: the single-electron tunneling current standard, the Josephson voltage standard, and the quantum Hall resistance standard. The name comes from a graphical description, see Fig. 9.17. The triangle sides, or "legs," are the equations governing the three quantum standards, and the vertices are given by the quantities involved in the experiments: voltage V, current I and frequency f. The achievement of quantum metrology triangle "closure," with a combined uncertainty in the 10^{-8} range, is the holy grail of primary electrical metrology.

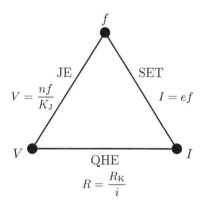

Figure 9.17
Quantum metrology triangle.

The original version of the quantum metrology triangle (Likharev and Zorin, 1985) is based on Ohm's law $V = RI$. V, I, and resistance R are determined by Josephson, SET pump, and quantum Hall equations, respectively (taking into account experimental constants, e.g., those related to voltage, current, or resistance ratio devices involved). If Ohm's law holds within the experimental uncertainty, the triangle "closes" and the mutual consistency of the three quantum experiments is verified. A tough experimental difficulty is the need to scale the SET pump current to the $100\,\mathrm{nA}-1\,\mu\mathrm{A}$ range, performed with cryogenic current comparators (Sec. 4.5.1) having a large turn ratio (>10000).

The ECCS experiment (Sec. 9.4.2) permits an alternative realization of the quantum metrology triangle. The capacitance determination of the ECCS involves Josephson and SET pump equations, and can be compared with measurements traceable to the quantum Hall effect (Sec. 9.3.1), or in alternative to the calculable capacitor.

9.5   The future of SI

The realization of electrical units in the present SI is mechanical in its essence; the base unit ampere is defined in terms of force and length (Sec. 9.1.3), and therefore is related to the present definition of the kilogram, which is in turn based on a material artfact (the international prototype) whose stability in time is presently questioned (Davis, 2003).

Electrical units in the dc and low frequency ranges are presently widely reproduced with Josephson and quantum Hall effect (and possibly, in the future, with electron counting capacitance experiments). Such quantum effects are strictly linked to the values of fundamental constants, which are believed to be universal and invariant quantities. Fundamental constants are presently measured in terms of SI units, and their value and uncertainty is the subject of redeterminations over time (Appendix C).

The realization of electrical units in general, and of impedance units in particular, would be significantly improved if the kilogram were redefined so as to be linked to an exact numerical value of h, and if the ampere were to be redefined so as to be linked to an exact numerical value of the elementary charge e.

The 24th CGPM in October 2011 has adopted a resolution on the possible future revision of the SI, by setting out a detailed road map toward the future changes. The proposal is to redefine four of the SI base units, namely, the kilogram, the ampere, the kelvin, and the mole, in terms of invariants of nature. The new definitions will be based on fixed numerical values of the Planck constant h, the elementary charge e, the Boltzmann constant k, and the Avogadro constant N_A, respectively (as a consequence, the Josephson and von Klitzing constants K_J and R_K will be also fixed). The detailed structure of the new SI is under discussion (new SI brochure).

QHE and ECCS experiments will become, in the new SI, unit realizations. In turn, experiments that are presently employed to realize capacitance and inductance units (the calculable capacitor and calculable inductor), in the new SI might be employed as ways to determine the electric and magnetic constants, no more having fixed values.

Appendix A

SI

For convenience, in the SI certain derived units have been given special names and symbols. There are 22 of such units: those related to electromagnetism are listed in Table A.1. These special names and symbols may themselves be used in combination with the names and symbols for based units and other derived units, to express the units of other derived quantities. For example, permittivity can be expressed in farad per meter $(F\,m^{-1})$ or, in terms of SI base units, in $m^{-3}\,kg^{-1}\,s^4\,A^{-2}$.

Table A.1
SI-derived units related to electromagnetism, having special names.

Derived quantity	name	symbol	expression in terms of base units
frequency	hertz	Hz	s^{-1}
energy	joule	J	$m^2\,kg\,s^{-2}$
power	watt	W	$m^2\,kg\,s^{-3}$
electric charge	coulomb	C	$s\,A$
electric potential difference	volt	V	$m^2\,kg\,s^{-3}\,A^{-1}$
electric capacitance	farad	F	$m^{-2}\,kg^{-1}\,s^{-4}\,A^2$
electric resistance	ohm	Ω	$m^2\,kg\,s^{-3}\,A^{-2}$
electric conductance	siemens	S	$m^{-2}\,kg^{-1}\,s^3\,A^2$
magnetic flux	weber	Wb	$m^2\,kg\,s^{-2}\,A^{-1}$
magnetic flux density	tesla	T	$kg\,s^{-2}\,A^{-1}$
inductance	henry	H	$m^2\,kg\,s^{-2}\,A^{-2}$

The SI adopts a series of prefix names and prefix symbols to form the names and symbols of the decimal multiples and submultiples of SI units, ranging from 10^{-24} to 10^{24}. Table A.2 lists all SI prefix names and symbols.

Table A.2
The SI prefixes.

name	symbol	factor	name	symbol	factor
yocto	y	10^{-24}	deca	da	10^{1}
zepto	z	10^{-21}	hecto	h	10^{2}
atto	a	10^{-18}	kilo	k	10^{3}
femto	f	10^{-15}	mega	M	10^{6}
pico	p	10^{-12}	giga	G	10^{9}
nano	n	10^{-9}	tera	T	10^{12}
micro	μ	10^{-6}	peta	P	10^{15}
milli	m	10^{-3}	exa	E	10^{18}
centi	c	10^{-2}	zetta	Z	10^{21}
deci	d	10^{-1}	yotta	Y	10^{24}

Appendix B

Harmonic electromagnetic fields

CONTENTS

In the following, Maxwell equations and Poynting theorem are written for the particular case of harmonic fields. The results are employed in Sec. 1.8 to link the concept of impedance to the physics of electromagnetic fields.

B.1   Harmonic fields

Let us consider the generic real-valued scalar or vector *field* $X(\boldsymbol{r}, t)$; in the sinuosidal steady state with angular frequency ω, it can be expressed in phasor form $X(\boldsymbol{r}, t) = \mathrm{Re}\left[X(\boldsymbol{r})\exp(\mathrm{j}\omega t)\right]$.

Time partial derivative of $X(\boldsymbol{r}, t)$ can be written as

$$\frac{\partial X(\boldsymbol{r}, t)}{\partial t} = \mathrm{Re}\left[\mathrm{j}\omega X(\boldsymbol{r}) \cdot \exp(\mathrm{j}\omega t)\right].$$

The scalar product between two fields $A(\boldsymbol{r}, t)$ and $B(\boldsymbol{r}, t)$ can be written in terms of their amplitudes as

$$
\begin{aligned}
A(\boldsymbol{r}, t) \cdot B(\boldsymbol{r}, t) &= \left(\frac{1}{2}\left[A(\boldsymbol{r})\exp(\mathrm{j}\omega t) + A^*(\boldsymbol{r})\exp(-\mathrm{j}\omega t)\right]\right) \\
&\quad \cdot \left(\frac{1}{2}\left[B(\boldsymbol{r})\exp(\mathrm{j}\omega t) + B^*(\boldsymbol{r})\exp(-\mathrm{j}\omega t)\right]\right) \\
&= \frac{1}{4}\left(A(\boldsymbol{r}) \cdot B^*(\boldsymbol{r}) + A^*(\boldsymbol{r}) \cdot B(\boldsymbol{r})\right) \\
&\quad + \frac{1}{4}\left(A(\boldsymbol{r}) \cdot B(\boldsymbol{r})\exp(2\mathrm{j}\omega t) + A^*(\boldsymbol{r}) \cdot B^*(\boldsymbol{r})\exp(-2\mathrm{j}\omega t)\right) \\
&= \frac{1}{2}\left[A^*(\boldsymbol{r}) \cdot B(\boldsymbol{r})\right] + \frac{1}{2}\mathrm{Re}\left[A(\boldsymbol{r}) \cdot B(\boldsymbol{r})\exp(2\mathrm{j}\omega t)\right].
\end{aligned}
$$

The product contains a time-independent term, and an harmonic term at frequency 2ω.

B.2 Maxwell equations for harmonic fields

In SI, Maxwell equations can be written in terms of vector fields E, D, B, H, and the charge ρ and current J densities as

$$\mathbf{div}\, D = \rho; \quad \mathbf{div}\, B = 0; \quad \mathbf{curl}\, E = -\frac{\partial B}{\partial t}; \quad \mathbf{curl}\, H = J + \frac{\partial D}{\partial t}. \quad (B.1)$$

For harmonic fields, Eq. (B.1) can be rewritten as

$$\mathbf{div}\, D = \rho; \quad \mathbf{div}\, B = 0; \quad \mathbf{curl}\, E = -\mathrm{j}\omega B; \quad \mathbf{curl}\, H = J + \mathrm{j}\omega D.$$

B.3 Electromagnetic fields and energy

Lorentz force F on a charge q, having velocity v, generated by an electromagnetic field (E, B) is

$$F = qE + qv \times B,$$

and its power P is

$$P = v \cdot F = qv \cdot E.$$

For a macroscopic system of currents (J), the electromagnetic power over a volume $\mathcal{V}$ is

$$P = \int_{\mathcal{V}} J(r, t) \cdot E(r, t)\, \mathrm{d}\mathcal{V},$$

which, for an harmonic field, can be written

$$P = \frac{1}{2} \int_{\mathcal{V}} J^*(r) \cdot E(r)\, \mathrm{d}\mathcal{V}. \quad (B.2)$$

Eq. (B.2) can be transformed, by applying in sequence the fourth Maxwell equation $(J = \mathbf{curl}\, H - \mathrm{j}\omega D \to J^* = \mathbf{curl}\, H^* + \mathrm{j}\omega D^*)$, a vector identity $(\mathbf{div}\, E \times H = E \cdot \mathbf{curl}\, H - H \cdot \mathbf{curl}\, E)$ and the third Maxwell equation $(\mathbf{curl}\, E = \mathrm{j}\omega B)$ into

$$P = \frac{1}{2} \int_{\mathcal{V}} [-\mathrm{j}\omega B \cdot H^* + \mathrm{j}\omega E \cdot D^* - \mathbf{div}\,(E \times H^*)]\, \mathrm{d}\mathcal{V}. \quad (B.3)$$

Let us introduce the electrical w_e and magnetic w_m field-energy densities

$$w_e(\boldsymbol{r}, t) = \frac{1}{2} \left[\boldsymbol{E}(\boldsymbol{r}, t) \cdot \boldsymbol{D}(\boldsymbol{r}, t) \right],$$

$$w_m(\boldsymbol{r}, t) = \frac{1}{2} \left[\boldsymbol{B}(\boldsymbol{r}, t) \cdot \boldsymbol{H}(\boldsymbol{r}, t) \right],$$

which for an harmonic field can be rewritten as

$$w_e(\boldsymbol{r}) = \frac{1}{2} \left[\boldsymbol{E}(\boldsymbol{r}) \cdot \boldsymbol{D}^*(\boldsymbol{r}) \right],$$

$$w_m(\boldsymbol{r}) = \frac{1}{2} \left[\boldsymbol{B}(\boldsymbol{r}) \cdot \boldsymbol{H}^*(\boldsymbol{r}) \right],$$

and the *Poynting vector* $\boldsymbol{S}(\boldsymbol{r}, t) = \boldsymbol{E}(\boldsymbol{r}, t) \cdot \boldsymbol{H}(\boldsymbol{r}, t)$, that for an harmonic field becomes $\boldsymbol{S}(\boldsymbol{r}) = \boldsymbol{E}(\boldsymbol{r}) \cdot \boldsymbol{H}^*(\boldsymbol{r})$, for which Gauss's theorem holds ($\mathcal{A}$ is the surface with normal $\boldsymbol{n}$ enclosing $\mathcal{V}$)

$$\int_{\mathcal{V}} \mathbf{div}\, \boldsymbol{S}(\boldsymbol{r}) \mathrm{d}\mathcal{V} = \int_{\mathcal{A}} \boldsymbol{S}(\boldsymbol{r}) \cdot \boldsymbol{n}(\boldsymbol{r}) \, \mathrm{d}\mathcal{A}$$

we can rewrite Eq. (B.3) in the *Poynting theorem*, the energy conservation theorem, for harmonic fields:

$$\frac{1}{2} \int_{\mathcal{V}} \boldsymbol{J}^*(\boldsymbol{r}) \cdot \boldsymbol{E}(\boldsymbol{r}) \mathrm{d}\mathcal{V} + 2\mathrm{j}\omega \int_{\mathcal{V}} (w_m - w_e) \, \mathrm{d}\mathcal{V} + \int_{\mathcal{A}} \boldsymbol{S}(\boldsymbol{r}) \cdot \boldsymbol{n}(\boldsymbol{r}) \, \mathrm{d}\mathcal{A} = 0, \quad \text{(B.4)}$$

where the following terms can be identified:

$\frac{1}{2} \int_{\mathcal{V}} \boldsymbol{J}^*(\boldsymbol{r}) \cdot \boldsymbol{E}(\boldsymbol{r}) \mathrm{d}\mathcal{V}$ is the mechanical work of the electric field on the electrical charges.

$2\mathrm{j}\omega \int_{\mathcal{V}} (w_m - w_e) \, \mathrm{d}\mathcal{V}$ is the energy of the electromagnetic field in the volume $\mathcal{V}$. If the quantity is completely imaginary (that is, if w_m and w_e are pure real quantities) the energy is stored in the field. An imaginary part of w_m and w_e is a dissipation (caused by magnetic or dielectric losses in the material).

$\int_{\mathcal{A}} \boldsymbol{S}(\boldsymbol{r}) \cdot \boldsymbol{n}(\boldsymbol{r}) \, \mathrm{d}\mathcal{A}$ is the electromagnetic radiant energy crossing the contour $\mathcal{A}$.

Appendix C

CODATA recommended values

The Committee on Data for Science and Technology (CODATA) is an interdisciplinary committee of the International Council for Science. The CODATA Task Group on Fundamental Constants has the purpose of periodically providing the international scientific and technological communities with an internationally accepted set of values of the fundamental physical constants and closely related conversion factors for use worldwide.

The CODATA 2010 set of recommended values is based on results of experiments published before 31 December 2010 and is published in international reviews (CODATA 2010) and on the NIST website.

Table C.1 gives a subset of the 2010 fundamental constant recommended values of interest for impedance metrology.

Table C.1
The 2010 CODATA recommended values of the fundamental physical constant, of interest for impedance measurement.

Quantity	symbol	numerical value	unit	Relative std. unc.
speed of light in vacuum	c, c_0	299 792 458	$m\,s^{-1}$	(exact)
magnetic constant	μ_0	$4\pi \times 10^{-7}$	$H\,m^{-1}$	(exact)
electric constant $1/(\mu_0 c^2)$	ϵ_0	$8.854\,187\,817\ldots \times 10^{-12}$	$F\,m^{-1}$	(exact)
characteristic impedance of vacuum $\mu_0 c$	Z_0	$376.730\,313\,461\ldots$	Ω	(exact)
Planck constant	h	$6.626\,069\,57(29) \times 10^{-34}$	$J\,s$	4.4×10^{-8}
elementary charge	e	$1.602\,176\,565(35) \times 10^{-19}$	C	2.2×10^{-8}
magnetic flux quantum $h/2e$	Φ_0	$2.067\,833\,758(46) \times 10^{-15}$	Wb	2.2×10^{-8}
conductance quantum $2e^2/h$	G_0	$7.748\,091\,734\,6(25) \times 10^{-5}$	S	3.2×10^{-10}
Josephson constant $2e/h$	K_J	$483\,597.870(11) \times 10^9$	$Hz\,V^{-1}$	2.2×10^{-8}
von Klitzing constant h/e^2	R_K	$25\,812.807\,443\,4(84)$	Ω	3.2×10^{-10}
conventional value of Josephson constant	K_{J-90}	$483\,597.9 \times 10^9$	$Hz\,V^{-1}$	(exact)
conventional value of von Klitzing constant	R_{K-90}	$25\,812.807$	Ω	(exact)

Appendix D

Reactance Chart

The *reactance chart*, or *reactance nomograph*, Fig. D.1, permit a quick graphical calculation of the reactance absolute value of a pure inductor or capacitor at a given frequency. The log-log format for the chart is used traditionally because it allows many decades to be plotted in a small area.

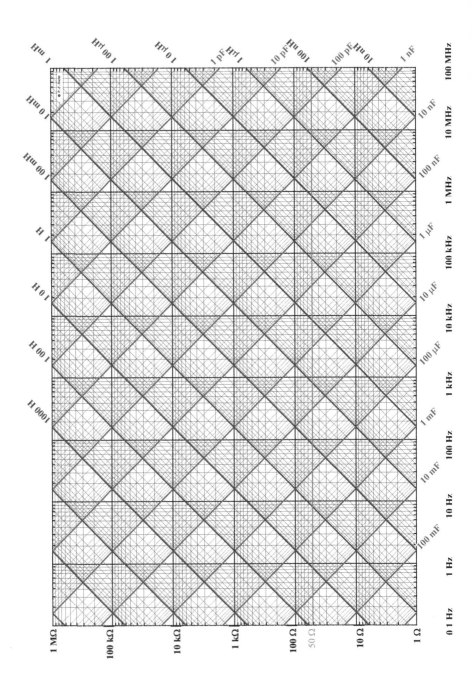

Nomenclature

BIPM	International Bureau of Weights and Measures (in French: *Bureau International des Poids et Mesures*) is an intergovernmental organization under the authority of the CGPM and the supervision of the CIPM. The BIPM acts in matters of world metrology, particularly concerning the demand for measurement standards of ever-increasing accuracy, range, and diversity, and the need to demonstrate equivalence between national measurement standards. It is one of the three organizations, with CGPM and CIPM, established to maintain the SI under the terms of the Convention du Mètre (Metre Convention) of 1875. BIPM is located in Sèvres, Paris, France.
CGPM	General Conference of Weights and Measures (in French: *Conférence Générale des Poids et Mesures*). It is one of the three organizations, with BIPM and CIPM, established to maintain the SI under the terms of the Convention du Mètre (Metre Convention) of 1875. It meets every four to six years at BIPM.
CIPM	International Committee for Weights and Measures (in French: *Comité International des Poids et Mesures*). It is one of the three organizations, with BIPM and CGPM, established to maintain the SI under the terms of the Convention du Mètre (Metre Convention) of 1875. CIPM is made up of eighteen individuals, each from a different Member State under the Metre Convention. Its principal task is to promote world wide uniformity in units of measurement, and it does this by direct action or by submitting draft resolutions to the CGPM. The CIPM meets annually at the BIPM and, among other matters, discusses reports presented to it by its Consultative Committees.
CMI	Czech Metrology Institute. The national metrology institute of the Czech Republic for electrical quantities.
CNRS	National Center of Scientific Research (in French: *Centre National de la Recherche Scientifique*), governmental research organization in France.
CODATA	Committee on Data for Science and Technology. The Task Group on Fundamental Constants periodically provide the in-

ternational scientific and technological communities with an internationally accepted set of values of the fundamental physical constants and closely related conversion factors for use worldwide.

CTU   Czech Technical University in Prague.

GUM   The *Guide to the Expression of Uncertainty in Measurement*, see References.

IEEE, IRE,   The Institute of Electrical and Electronics Engineers (read I-
AIEE   triple-E) is a nonprofit professional association dedicated to advancing technological innovation related to electricity, based in the USA. It was formed in 1963 by the merger of the Institute of Radio Engineers (IRE) and the American Institute of Electrical Engineers (AIEE). IEEE publishes well over 100 peer-reviewed journals. The IEEE Standards Association is in charge of the standardization activities of the IEEE.

IET, IEE,   The Institution of Engineering and Technology (IET) is a
IIE   British professional body for those working in engineering and technology in the United Kingdom and worldwide. It was formed in 2006 by the merger of the Institution of Electrical Engineers (IEE) and the Institution of Incorporated Engineers (IIE).

INRIM   Istituto Nazionale di Ricerca Metrologica, Torino, Italy. The national metrology institute of Italy for electrical quantities.

JCGM   Joint Committee for Guides in Metrology of the CGPM.

LNE   Laboratoire National de Métrologie et d'Essais. The national metrology institute of France for electrical quantities.

NIST   National Institute of Standards and Technology. The national metrology institute of the United States of America for electrical quantities.

NMIA   National Measurement Institute, Australia. The national metrology institute of Australia for electrical quantities.

NMIA   National Metrology Institute of Australia for electrical quantities.

NPL   National Physical Laboratory, Teddington, UK. The national metrology institute of the United Kingdom for electrical quantities.

NRC-INMS   National Research Council Institute for National Measurement Standards, Canada. The national metrology institute of Canada for electrical quantities.

PTB   Physikalisch-Technische Bundesanstalt, Braunschweig, Germany. The national metrology institute of Germany for electrical quantities.

QHE   Quantum Hall effect.

SI   International System of Units (in French: *Système International d'Unités*).

VIM   The *International Vocabulary of Metrology Basic and General Concepts and Associated Terms*, see References.

Bibliography

Abbott B., Davis D. R., Phillips N. J., and Eshraghian K., "Simple derivation of the thermal noise formula using window-limited Fourier transforms and other conundrums," *IEEE Trans. Educ.*, vol. 39, n. 1, pp. 1–13, Feb. 1996.

AD5934: "250 kSPS, 12-Bit Impedance Converter, Network Analyzer," Analog Devices Datasheet.

Agilent 1287: "Specifying calibration standards and kits for Agilent vector network analyzers," Agilent Tech., Appl. Note 1287-11, 5989-4840EN.

Agilent 346-3: "Effective impedance measurement using OPEN/SHORT/LOAD correction," Agilent Tech., Appl. Note 346-3, 1998.

Ahlers F. J., Jeanneret B., Overney F., Schurr J., and Wood B. M., "Compendium for precise ac measurements of the quantum Hall resistance," *Metrologia*, vol. 46, no. 5, pp. R1–R11, Oct. 2009.

Amato G. and Enrico E., *Superconductivity – Theory and Applications.* In-Tech, 2011, ch. 13: Current status and technological limitations of hybrid superconducting-normal single electron transistors, pp. 279–300, ISBN: 978-953-307-151-0.

Anderson K. F., "Constant current loop impedance measuring system that is immune to the effects of parasitic impedances," US Patent 5 731 469, Dec. 6, 1994.

Anderson K. F., "NASA's Anderson loop," *IEEE Instrum. Meas. Mag.*, vol. 1, no. 1, pp. 5–30, 1998.

Aoki T. and Yokoi K., "Capacitance scaling system," *IEEE Trans. Instr. Meas.*, vol. 46, n. 2, pp. 474–476, Apr. 1997.

Aoki T., Suzuki K., and Yokoi K., "Calibration method for four-terminal-pair capacitance standards: Progress report," in *Conference on Precision Electromagnetic Measurements CPEM 1998 Digest*, Washington DC, USA, pp. 36–37, Jul. 6–10, 1998.

Arri E. and Noce G., "A high accuracy self-calibrating bridge with coupled inductive ratio arms used for standard inductors comparison," *IEEE Trans. Instr. Meas*, vol. 1M-23, no. 1, pp. 72–79, Mar. 1974.

Astbury N. F., "The design, construction and use of resistors with calculable performance," *J. IEE*, vol. 76, no. 460, pp. 389–396, 1935.

Astin A. V., "Nature of energy losses in air capacitors at low frequencies," *J. Res. Nat. Bur. Stand.*, vol. 22, pp. 673–695, Jun. 1939.

ASTM E 1004-09: "Standard test method for determining electrical conductivity using the electromagnetic (eddy-current) method," 2009.

Averin D. V. and Likharev K. K., "Coulomb blockade of single-electron tunneling, and coherent oscillations in small tunnel junctions," *J. Low Temp. Phys.*, vol. 62, no. 3-4, pp. 345–373, 1986.

Avramov-Zamurovic S., Koffman A. D., Oldham N. M., and Waltrip B. C., "The sensitivity of a method to predict a capacitor's frequency characteristic," *IEEE Trans. Instr. Meas*, vol. 49, no. 2, pp. 398–404, Apr. 2000.

Avramov-Zamurovic S., Koffman A. D., Waltrip B. C., and Wang Y., "Evaluation of a capacitance scaling system," *IEEE Trans. Instr. Meas.*, vol. 56, no. 6, pp. 2160–2163, Dec. 2007.

Awan S., Kibble B., and Schurr J., *Coaxial Electrical Circuits for Interference – Free Measurements*, ser. IET Electrical Measurement. Institute of Engineering and Technology, 2011, ISBN: 9781849190695.

Awan S., Kibble B., and Robinson I., "Calibration of IVDs at frequencies up to 1 MHz by permuting capacitors," *IEE Proc. Science, Meas. and Tech.*, vol. 147, no. 4, pp. 193–195, Jul. 2000.

Awan S. A. and Kibble B. P., "Towards accurate measurement of the frequency dependence of capacitance and resistance standards up to 10 MHz," *IEEE Trans. Instr. Meas.*, vol. 54, no. 2, pp. 516–520, Apr. 2005.

Awan S. A. and Kibble B. P., "A universal geometry for calculable frequency-response coefficient of LCR standards and new 10-MHz resistance and 1.6-MHz quadrature-bridge systems," *IEEE Trans. Instr. Meas.*, vol. 56, no. 2, pp. 221–225, Apr. 2007.

Awan S. A., Callegaro L., and Kibble B. P., "Resonance frequency of four terminal-pair air-dielectric capacitance standards and closing the metrological impedance triangle," *Meas. Sci. Technol.*, vol. 15, pp. 969–972, 2004.

Bachmair H. and Vollmert R., "Comparison of admittances by means of a digital double-sinewave generator," *IEEE Trans. Instr. Meas.*, vol. IM-29, no. 4, pp. 370–372, Dec. 1980.

Baker-Jarvis J., Janezic M. D., Grosvenor J. H., and Geyer R. G., "Transmission/reflection and short-circuit line methods for measuring permittivity and permeability," National Institute of Standard and Technology, Tech. Note 1355-R, Dec. 1993, 236 pp.

Bao J.-Z., Davis C. C., and Schmukler R. E., "Impedance spectroscopy of human erhthrocytes: system calibration and nonlinear modeling," *IEEE Trans. Biomed. Eng.*, vol. 40, no. 4, pp. 364–378, Apr. 1993.

Barkhausen H., "Zwei mit Hilfe der neuen Verstärker entdeckte Erscheinungen," *Phys. Z*, vol. 20, p. 401–403, 1919.

Barsoukov E. and Macdonald J., *Impedance Spectroscopy: Theory, Experiment, and Applications*, 2nd ed. Hoboken, NJ, USA: J. Wiley & Sons, Inc., 2005, ISBN: 9780471647492.

Baxter R., "RMS sensor techniques in modern AC measuring instruments," in *Developments of AC Voltage Measurements, IEE Colloquium on*, pp. 4/1–4/5, Apr. 1992.

Bechhoefer J., "Kramers–Krönig, Bode, and the meaning of zero," *Am. J. Phys.*, vol. 79, no. 10, pp. 1053–1059, 2011.

Bennett W. R., "Spectra of quantized signals," *Bell Syst. Tech. J*, vol. 27, no. 3, pp. 446–472, 1948.

Berggren C., Bjarnason B., and Johansson G., "Capacitive biosensors," *Electroanalysis*, vol. 13, no. 3, pp. 173–180, 2001.

Bianco B., Corana A., Gogioso L., Ridella S., and Parodi M., "Open-circuited coaxial lines as standards for microwave measurements," *Electron. Lett.*, vol. 16, no. 10, pp. 373–374, 1980.

Bilau T. Z., Megyeri T., Sarhegyi A., Markus J., and Kollar I., "Four-parameter fitting of sine wave testing result: iteration and convergence," *Computer Std. and Interf.*, vol. 26, pp. 51–56, 2003.

Binnie A. J. and Foord T. R., "Leakage inductance and interwinding capacitance in toroidal ratio transformers," *IEEE Trans. Instr. Meas.*, vol. IM-16, no. 4, pp. 307–314, Dec. 1967.

Blumlein A. D., "Alternating current bridge circuits," British Patent, 1928, 323037.

Blumlein A. D., "Improvements in or relating to electrical bridge arrangements," UK Patent GB581 164, 1946.

Bohacek J., "A QHE-based system for calibrating impedance standards," *IEEE Trans. Instr. Meas.*, vol. 53, no. 4, pp. 977–980, Aug. 2004.

Bohacek J. and Wood B. M., "Octofilar resistors with calculable frequency dependence," *Metrologia*, vol. 38, pp. 241–247, 2001.

Bohacek J., Jursa T., and Sedlacek R., "Calibrating standards of mutual inductance," in *Instrumentation and Measurement Technology Conference IMTC 2005 Proc.*, Canada, Ottawa, 17–19, vol. 2, May 2005, pp. 1002–1004.

Bonaldi M., Falferi P., Dolesi R., Cerdonio M., and Vitale S., "High Q tunable LC resonator operating at cryogenic temperature," *Rev. Sci. Instrum.*, vol. 69, no. 10, pp. 3690–3694, 1998.

Bounouh A. and Bélières D., "New approach in ac voltage references based on micronanosystems," *Metrologia*, vol. 48, no. 1, p. 40–46, 2011.

Bounouh A., Satrapinski A., Ziade F., Morilhat A., and Leprat D., "Direct comparisons of ac resistance standards of various technology designs," in *Conference on Precision Electromagnetic Measurements CPEM 2008 Digest*, Broomfield, CO, USA, pp. 550–551, Jun. 8–13, 2008.

Bounouh A., "Numerical computations and measurements on calculable resistance standards based on NiCr thin films," in *Conference on Precision Electromagnetic Measurement CPEM 2004 Digest*, London, UK, pp. 370–371, Jun. 2, – July 2, 2004.

Bounouh A., "Fabrication and characterisation of thin film coaxial ac/dc resistors for the determination of R_K," in *IMEKO XVIII World Congress Proc.*, Rio de Janeiro, Brazil, p. 00572, Sep. 17–22, 2006.

Bowler N. and Huang Y., "Electrical conductivity measurement of metal plates using broadband eddy-current and four-point methods," *Meas. Sci. Technol.*, vol. 16, no. 11, p. 2193–2200, 2005.

Briggs M. E., Gammon R. W., and Shaumeyer J. N., "Measurement of the temperature coefficient of ratio transformers," *Rev. Sci. Instrum.*, vol. 64, no. 3, pp. 756–759, 1993.

Brinkmann F., Dam N. E., Deák E., Durbiano F., Ferrara E., Fükö J., Jensen H. D., Máriássy M., Shreiner R. H., Spitzer P., Sudmeier U., Surdu M., and Vyskočil L., "Primary methods for the measurement of electrolytic conductivity," *Accred. Qual. Assur.*, vol. 8, pp. 346–353, 2003.

Buckley T. J., Hamelin J., and Moldover M. R., "Toroidal cross capacitor for measuring the dielectric constant of gases," *Rev. Sci. Instrum.*, vol. 71, no. 7, pp. 2914–2921, 2000.

Butterworth S., "Capacity and eddy current effects in inductometers," *Proc. Physical Society London*, vol. 33, no. 1, p. 312–354, 1920.

Büttiker M., "Scattering theory of thermal and excess noise in open conductors," *Phys. Rev. Lett.*, vol. 65, no. 23, pp. 2901–2904, 1990.

Cabiati F. and Bosco G. C., "CCE intercomparison of 10 mH inductors: measurement method and results at IEN," CCE, Tech. Rep. 92-52, 1992, working document.

Cabiati F. and D'Elia V., "High-accuracy voltage and current transmission by a four-coaxial cable," in *Conference on Precision Electromagnetic Measurements CPEM 2000 Digest*, Sydney, NSW, Australia, pp. 435–436, May 14–19, 2000.

Cabiati F. and D'Elia V., "A new architecture for high-accuracy admittance measuring systems," in *Conference on Precision Electromagnetic Measurements CPEM 2002 Digest*, Ottawa, Canada, pp. 178–179, Jun. 16–21, 2002.

Cabiati F. and D'Emilio S., "Low frequency transmission errors in multi-coaxial cables and four-port admittance standard definition," *Alta Frequenza*, vol. XLIV, pp. 609–616 (319E–326E), Oct. 1975.

Cabiati F. and La Paglia G., "Load effect reduction in low frequency ratio devices by an active synchronous technique," in *EUROMEAS-77*. Sussex, UK: IEE, pp. 50–52, Sep. 5–9, 1977.

Cabiati F., Bosco G. C., and Sosso A., "Impedance comparison in the low-medium range through precision voltage measurements," in *XIII IMEKO World Congr. Proc.*, vol. 1, Torino, pp. 335–339, Sep. 5–9, 1994.

Cabiati F. and Pogliano U., "High-accuracy two-phase digital generator with automatic ratio and phase control," *IEEE Trans. Instr. Meas.*, vol. IM-36, no. 2, pp. 411–417, Jun. 1987.

Callegaro L., "Calibration of impedances by the substitution method: numerical uncertainty evaluation," in *XI IMEKO TC-4 Symp. Proc.*, Lisbon, Portugal, pp. 481–484, Sep. 13–14, 2001.

Callegaro L. and D'Elia V., "Automated system for inductance realization traceable to AC resistance with a three-voltmeter method," *IEEE Trans. Instr. Meas.*, vol. 50, pp. 1630–1633, Dec. 2001.

Callegaro L., Bosco G., Capra P., and Serazio D., "A remotely controlled coaxial switch for impedance standard calibration," *IEEE Trans. Instr. Meas.*, vol. 51, no. 4, pp. 628–631, Aug. 2002.

Callegaro L., Bosco G., D'Elia V., and Serazio D., "Direct-reading absolute calibration of ac voltage ratio standards," *IEEE Trans. Instr. Meas.*, vol. 52, no. 2, pp. 380–383, Apr. 2003.

Callegaro L., D'Elia V., and Serazio D., "10-nF capacitance transfer standard," *IEEE Trans. Instr. Meas.*, vol. 54, no. 5, pp. 1869–1872, Oct. 2005.

Callegaro L., D'Elia V., and Bohacek J., "Four-terminal-pair inductance comparison between INRIM and CTU," *IEEE Trans. Instr. Meas.*, vol. 58, no. 1, pp. 87–93, Jan. 2009.

Callegaro L., D'Elia V., and Gasparotto E., "Impedance comparison at power frequency by asynchronous sampling," in *Conference on Precision Electromagnetic Measurements CPEM 2010 Digest*, Daejeon, Korea, pp. 322–323, Jun. 13-18, 2010.

Callegaro L., "S-matrix method for high-frequency calibration of capacitors: uncertainty evaluation," in *Conference on Precision Electromagnetic Measurements CPEM 2006 Digest*, Torino, Italy, pp. 536–537, July 9–14, 2006.

Callegaro L., "EUROMET.EM-S20: Intercomparison of a 100 mH inductance standard (Euromet Project 607)," *Metrologia*, vol. 44, no. 1A, p. 01002, 2007.

Callegaro L., "The metrology of electrical impedance at high frequency: a review," *Meas. Sci. Technol.*, vol. 20, no. 2, p. 022002, 2009.

Callegaro L. and D'Elia V., "Automatic compensation technique for alternating current metrology based on synchronous filtering," *Rev. Sci. Instrum.*, vol. 69, no. 12, pp. 4238–4241, 1998.

Callegaro L. and Durbiano F., "Four terminal-pair impedances and scattering parameters," *Meas. Sci. Technol.*, vol. 14, pp. 523–529, 2003.

Callegaro L., Cassiago C., and D'Elia V., "Optical linking of reference channel for lock-in amplifiers," *Meas. Sci. Technol.*, vol. 10, no. 7, p. N91–N92, 1999.

Callegaro L., D'Elia V., and Trinchera B., "Realization of the farad from the dc quantum Hall effect with digitally assisted impedance bridges," *Metrologia*, vol. 47, no. 4, p. 464–472, 2010.

Callen H. B. and Welton T. A., "Irreversibility and generalized noise," *Phys. Rev.*, vol. 83, no. 1, pp. 34–40, Jul. 1951.

Campbell A., "On a standard of mutual inductance," *Proc. Royal Soc. of London A*, vol. 79, no. 532, pp. 428–435, 1907.

Campbell A. and Childs E. C., *The Measurement of Inductance, Capacitance, and Frequency.* Macmillan and Co., Ltd., London, 1935.

Campbell G. A. and Foster R. M., "Maximum output networks for telephone substation and repeater circuits," *AIEE Trans.*, vol. 39, no. 1, pp. 231–290, Jan. 1920.

Carlin H., "The scattering matrix in network theory," *IRE Trans. on Circuit Theory*, vol. 3, no. 2, pp. 88–97, Jun. 1956.

Chen K.-F., "On the condition of four-parameter sine wave fitting," *Computer Std. and Interf.*, vol. 29, pp. 174–183, 2007.

Chen K.-F. and Xue Y.-M., "Four-parameter sine wave fitting by Gram–Schmidt orthogonalization," *Measurement*, vol. 41, pp. 76–87, 2008.

Christie S. H., "Experimental determination of the laws of magnetoelectric induction," *Phil. Trans. Roy. Soc.*, vol. 123, pp. 95–142, 1833.

Chua L. O., "Memristor – the missing circuit element," *IEEE Trans. on Circuit Theory*, vol. 18, pp. 507–519, Sep. 1971.

Chua L. O., Desoer C. A., and Kuh E. S., *Linear and Nonlinear Circuits.* McGraw-Hill, New York, 1987, ISBN: 0070108986.

CODATA 2010: "CODATA recommended values of the fundamental physical constants: 2010," *in press*, 2012. Also on ArXiv: 1203.5425 (Mar. 24, 2012).

Cole K. S. and Cole R. H., "Dispersion and absorption in dielectrics. I. Alternating current characteristics," *J. Chem. Phys.*, vol. 9, pp. 341–351, 1941.

Cole K. S. and Cole R. H., "Dispersion and absorption in dielectrics. II. Direct current characteristics," *J. Chem. Phys.*, vol. 10, pp. 98–105, 1942.

Corney A. C., "Digital generator–assisted impedance bridge," *IEEE Trans. Instr. Meas.*, vol. 52, no. 2, pp. 388–391, Apr. 2003.

Corney A. C., "A universal four-pair impedance bridge," *IEEE Trans. Instr. Meas.*, vol. IM-28, no. 3, pp. 211–215, Sep. 1979.

Côté M., "A four-terminal coaxial-pair Maxwell-Wien bridge for the measurement of self-inductance," *IEEE Trans. Instr. Meas.*, vol. 58, no. 4, pp. 962–966, Apr. 2009.

Cutkosky R. D., "Evaluation of the NBS unit of resistance based on a computable capacitor," *J. Res. Nat. Bur. Stand. A: Phys. Chem.*, vol. 65A, pp. 147–158, May–Jun. 1961.

Cutkosky R. D., "Four-terminal-pair networks as precision admittance and impedance standards," *Commun. Electron.*, vol. 70, pp. 19–22, Jan. 1964.

Cutkosky R. D., "Techniques for comparing four-terminal-pair admittance standards," *J. Res. Nat. Bur. Stand*, vol. 74C, no. 3, pp. 475–489, Jul.-Dec. 1970.

Cutkosky R. D. and Lee L. H., "Improved ten-picofarad fused silica dielectric capacitor," *J. Res. Nat. Bur. Std.*, vol. 69C, no. 3, pp. 173–179, Sep. 1965.

Cutkosky R. D. and Shields J. Q., "The precision measurement of transformer ratios," *IRE Trans. Instrum.*, vol. I-9, no. 2, pp. 243–250, Sep. 1960.

Damle B., Regier C. G., and Borisch S., "Fast and accurate AC RMS and DC measurement," USA Patent 7 016 796, Mar 21, 2006.

Daniel M. G., "Characterization of a fused silica capacitance standard," Sandia Nat. Lab., Albuquerque, NM 871 85-0665, Tech. Rep. SAND94-3146, Jan. 1995.

Davis R., "The SI unit of mass," *Metrologia*, vol. 40, no. 6, p. 299–305, 2003.

Deacon T. and Hill J., "Two-stage inductive voltage dividers," *Proc. IEE*, vol. 115, no. 6, pp. 888–892, Jun. 1968.

Delahaye F., "Technical guidelines for reliable measurements of the quantized Hall resistance," *Metrologia*, vol. 26, no. 1, p. 63–68, 1989.

Delahaye F. and Jeckelmann B., "Revised technical guidelines for reliable dc measurements of the quantized Hall resistance," *Metrologia*, vol. 40, no. 5, p. 217–223, 2003.

Dodd C. V. and Deeds W. E., "Analytical solutions to eddy-current probe-coil problems," *J. Appl. Phys.*, vol. 39, no. 6, pp. 2829–2838, 1968.

van Drieënhuizen B. P. and Wolffenbuttel R. F., "Integrated micromachined electrostatic true RMS-to-DC converter," *IEEE Trans. Instr. Meas.*, vol. 44, no. 2, pp. 370–373, 1995.

Dunn A. F., "Determination of an absolute scale of capacitance," *Canadian Journal of Phys.*, vol. 42, pp. 53–69, Jan. 1964.

Durand M., Lawall J., and Wang Y., "Fabry-Perot displacement interferometry for next-generation calculable capacitor," in *Conference on Precision Electromagnetic Measurements CPEM 2010 Digest*, Daejeon, Korea, pp. 111–112, Jun. 13-18, 2010.

Durin G. and Zapperi S., in *The Science of Hysteresis*. Eds. Bertotti G., Mayergoyz I. Amsterdam: Elsevier, 2005, ISBN: 9780124808744, vol. II, ch. The Barkhausen effect, pp. 181–267, also in arXiv:cond-mat/0404512.

Eaton W. P. and Smith J. H., "Micromachined pressure sensors: review and recent developments," *Smart Materials and Structures*, vol. 6, no. 5, p. 530–539, 1997.

Eckardt H., Behnke H.-G., Bemme W., Semyonov Y., and Shvedov O., "Precision computer-controlled decade capacitor," *IEEE Trans. Instr. Meas.*, vol. 48, no. 2, pp. 360–364, Apr. 1999.

Eklund G., "Absolute determination of loss angle," in *Conference on Precision Electromagnetic Measurements CPEM 1996 Digest*, Braunschweig, Germany, pp. 390–391, Jun. 17-20, 1996.

Elmquist R. E., "Calculable coaxial resistors for precision measurements," *IEEE Trans. Instr. Meas.*, vol. 49, no. 2, pp. 210–214, Apr. 2000.

Engen G., "The six-port reflectometer: an alternative network analyzer," *IEEE Trans. Microwave Theor. Tech.*, vol. MTT-25, no. 12, pp. 1075–1080, Dec. 1977.

Engen G., "An improved circuit for implementing the six-port technique of microwave measurements," *IEEE Trans. Microwave Theor. Tech.*, vol. MTT-25, no. 12, pp. 1080–1083, Dec. 1977.

Espenschied L. and Affel H. A., "Concentric conducting system," NY, USA Patent 1 835 031, Dec. 8, 1931.

Essen L. and Froome K. D., "The refractive indices and dielectric constants of air and its principal constituents at 24,000 Mc/s," *Proc. Phys. Soc. B*, vol. 64, no. 10, p. 862–865, 1951.

Feldman Y., Andrianov A., Polygalov E., Ermolina I., Romanychev G., Zuev Y., and Milgotin B., "Time domain dielectric spectroscopy: An advanced measuring system," *Rev. Sci. Instrum.*, vol. 67, no. 9, pp. 3208–3216, Sep. 1996.

Fellmuth B., Bothe H., Haft N., and Melcher J., "High-precision capacitance bridge for dielectric-constant gas thermometry," *IEEE Trans. Instr. Meas.*, vol. 60, no. 7, pp. 2522–2526, Jul. 2011.

Fellmuth B., Fischer J., Gaiser C., Jusko O., Priruenrom T., Sabuga W., and Zandt T., "Determination of the Boltzmann constant by dielectric-constant gas thermometry," *Metrologia*, vol. 48, no. 5, p. 382–390, 2011.

Fellner-Feldegg H., "Measurement of dielectrics in the time domain," *J. Phys. Chem.*, vol. 73, no. 3, pp. 616–623, 1969.

Ferguson J. G., "Classification of bridge methods of measuring impedances," *AIEE Trans.*, vol. 52, no. 3, pp. 861–867, Sep. 1933.

Fiander J. R. and Small G. W., "Effect of rotational skew on a symmetrical cross capacitor," *Metrologia*, vol. 45, no. 3, p. 362–367, 2008.

Fiebiger A. and Dröge K., "Bestimmung der Temperaturkoeffizienten von Induktivitätsnormalen des typs GR 1482," *PTB-Mitteilungen*, vol. 94, no. 1, pp. 10–13, 1984.

Fiorillo F., "Measurements of magnetic materials," *Metrologia*, vol. 47, pp. S114–S142, 2010.

Fluke 5500: *Fluke Mod. 5500A Multi-Product Calibrator Service Manual*, Rev. 6, 7/06 ed., Fluke Corp., Aug. 1995.

Fonseca da Silva M., Ramos P. M., and Cruz Serra A., "A new four parameter sine fitting technique," *Measurement*, vol. 35, pp. 131–137, 2004.

Foord T., Langlands R., and Binnie A., "Transformer-ratio bridge network with precise lead compensation and its application to the measurement of temperature and temperature difference," *Proc. IEEE*, vol. 110, no. 9, pp. 1693–1700, Sep. 1963.

Free G. M. and Jones R. N., "Calibration service for low-loss, three-terminal capacitance standards at 100 kHz and 1 MHz," NIST, Tech. Note 1348, Feb. 1992.

Frenkel R., "A superconductor analogue of the Thompson-Lampard theorem of electrostatics and its possible application to a new SI standard of dc resistance," *Metrologia*, vol. 30, pp. 117–132, 1993.

Fujiki H., Domae A., and Nakamura Y., "Analysis of time constant of calculable ac/dc resistors for the phase angle standard," *Jap. J. Appl. Phys.*, vol. 42, no. 8, Part 1, pp. 5357–5360, 2003.

Fulton T. A. and Dolan G. J., "Observation of single-electron charging effects in small tunnel junctions," *Phys. Rev. Lett.*, vol. 59, no. 1, pp. 109–112, Jul. 1987.

Garcia-Valenzuela A. and Guadarrama-Santana A., "Isotropic scaling law of the electrical capacitance," *Am. J. Phys.*, vol. 78, no. 12, pp. 1376–1378, 2010.

Gibbings D. L. H., "An alternating-current analogue of the Kelvin double bridge," *Proc. IEE – Part C: Monogr.*, vol. 109, no. 16, pp. 307–316, Sep. 1962.

Gibbings D. L. H., "A design for resistors of calculable a.c./d.c. resistance ratio," *Proc. IEE.*, vol. 110, pp. 335–347, 1963.

Goldfarb R. and Bussey H., "Method for measuring complex permeability at radio frequencies," *Rev. Sci. Instrum.*, vol. 58, no. 4, pp. 624–627, 1987.

Gottardi L., de Waard A., Usenko O., Frossati G., Podt M., Flokstra J., Bassan M., Fafone V., Minenkov Y., and Rocchi A., "Sensitivity of the spherical gravitational wave detector MiniGRAIL operating at 5 K," *Phys. Rev. D*, vol. 76, p. 102005, Nov. 2007.

Gregory A. P. and Clarke R. N., "Traceable measurements of the static permittivity of dielectric reference liquids over the temperature range 5-50 °C," *Meas. Sci. Tech.*, vol. 16, pp. 1506–1516, 2005.

Grohmann K. and Hechtfischer D., "Self-calibrating cryo current comparators for AC applications," *IEEE Trans. Instr. Meas.*, vol. IM-33, pp. 91–96, Jun. 1984.

Gugan D., "The use of eddy currents for measurements of bulk modulus: application to potassium," *Physica Scripta*, vol. 56, no. 1, p. 95–111, 1997.

Gulmez Y., Gulmez G., Turhan E., Ozkan T., Cinar M., and Sozen L., "A new design for calculable resistor," in *Conference on Precision Electromagnetic Measurements CPEM 2002 Digest*, Ottawa, Canada, pp. 348–349, Jun. 16–21, 2002.

GUM: "JCGM 100:2008, Evaluation of measurement data – Guide to the expression of uncertainty in measurement," Available online at www.bipm.org., 2008.

GUM Suppl. 1: "JCGM 101:2008, Evaluation of measurement data – Supplement 1 to the Guide to the expression of uncertainty in measurement – Propagation of distributions using a Monte Carlo method," Available online at www.bipm.org, 2008.

GUM Suppl. 2: "JCGM 102:2011, Evaluation of measurement data – Supplement 2 to the "Guide to the expression of uncertainty in measurement" – Extension to any number of output quantities," Available online at www.bipm.org., 2011.

Haddad R. J., "A resistor calculable from DC to $\omega = 10^5$ rad/s," Sch. Eng. Appl. Sci., George Washington Univ., M. S. Thesis, Apr. 1969.

Hague B., *Instrument Transformers: Their Theory, Characteristics and Testing*. London: I. Pitman and Sons, Ltd., 1936.

Hague B., *Alternating Current Bridge Methods*, 6th ed. London, UK: Pitman Publishing Ltd, 1971, revised by T. R. Foord and S. T. Mackay, ISBN 0273402919.

Hall H. P., "Four-terminal equal-power transformer-ratio-arm bridge," *IEEE Trans. Instr. Meas.*, vol. IM–19, no. 4, pp. 308–311, Nov. 1970.

Hall H. P., "How electronics changed impedance measurements," in *Instrumentation and Measurement Technology Conference IMTC2004 Proc.*, Como, Italy, pp. 2–8, May 18–20, 2004.

Händel P., "Properties of the IEEE-STD-1057 four-parameter sine wave fit algorithm," *IEEE Trans. Instr. Meas*, vol. 49, no. 6, pp. 1189–1193, Dec. 2000.

Hanke R., "An improved straddling method with triaxial guards for the calibration of inductive voltage dividers at 1592 Hz," *IEEE Trans. Instr. Meas.*, vol. 38, no. 5, pp. 974–978, Oct. 1989.

Hanke R. and Dröge K., "Bestimmung des Frequenzganges von Luftkondensatoren des Typs GR 1403 im Frquenzbereich von 10 kHz bis 1 MHz," *PTB-Mittilungen*, vol. 97, pp. 377–383, 1987 (in German).

Hanke R. and Ramm G., "Reduction of kelvin ratio arm errors in an ac double bridge using electronic control," *IEEE Trans. Instr. Meas.*, vol. 32, no. 1, pp. 145–146, Mar. 1983.

Hanke R., Koelling A., and Melcher J., "Inductance calibration in the frequency range from 50 Hz to 1 MHz at PTB," in *Conference on Precision Electromagnetic Measurements CPEM 2002 Digest*, Ottawa, Canada, pp. 186–187, Jun. 16–21, 2002.

Hanke R., "Precise kelvin double bridge for measuring dissipation factors and capacitances up to 1 F," *IEEE Trans. Instr. Meas.*, vol. IM-27, no. 4, pp. 434–436, Dec. 1978.

Harris I. A. and Spinney R. E., "The realization of high-frequency impedance standards using air-spaced coaxial lines," *IEEE Trans. Instr. Meas.*, vol. 13, no. 4, pp. 265–272, Dec. 1964.

Harrison P. and Rayner G., "A primary standard of mutual inductance," *Metrologia*, vol. 3, no. 1, pp. 1–12, 1967.

Hartshorn L., "The properties of mutual inductance standards at telephonic frequencies," *Proc. Physical Society London*, vol. 38, no. 1, p. 302–320, 1925.

Harvey I. K., "A precise low temperature dc ratio transformer," *Rev. Sci. Instrum*, vol. 43, no. 11, pp. 1626–1629, Nov. 1972.

Hauser M., Kraus L., and Ripka P., "Giant magnetoimpedance sensors," *IEEE Instrum. Meas. Mag.*, vol. 4, no. 2, pp. 28–32, Jun. 2001.

Helbach W. and Schollmeyer H., "Impedance measuring methods based on multiple digital generators," *IEEE Trans. Instr. Meas.*, vol. IM-36, pp. 400–405, Jun. 1987.

Helbach W., Marczinowski P., and Trenkler G., "High-precision automatic digital AC bridge," *IEEE Trans. Instr. Meas.*, vol. IM-32, no. 1, pp. 159–162, Mar. 1983.

Herlufsen H., "Dual-channel FFT analysis (part I & II)," *Brüel & Kjær Technical Review*, no. 1 and 2, pp. 3–56, 3–45, 1984.

Hersh J. F., "New standard inductors – more terminals, less inductance," *General Radio Experim.*, vol. 34, no. 10, pp. 6–9, Oct. 1960.

Hess D. T. and Clarke K. K., "Circuit techniques for use in a digital phase-angle generator," *IEEE Trans. Instr. Meas.*, vol. IM-36, no. 2, pp. 394–399, Jun. 1987.

Hill J. J. and Deacon T. A., "Voltage-ratio measurement with a precision of parts in 10^9 and performance of inductive voltage dividers," *IEEE Trans. Instr. Meas.*, vol. IM-17, no. 4, pp. 269–278, Dec. 1968.

Hill J. and Deacon T., "Theory, design and measurement of inductive voltage dividers," *Proc. IEE*, vol. 115, no. 5, pp. 727–735, May 1968.

Hill J. J. and Miller A., "A seven-decade adjustable-ratio inductively-coupled voltage divider with 0.1 part per million accuracy," *IEE Proc. B: Electron. Commun. Eng.*, vol. 109, no. 44, pp. 157–162, Mar. 1962.

Hoer C. A., "The six-port coupler: a new approach to measuring voltages, current, power, impedance and phase," *IEEE Trans. Instr. Meas.*, vol. IM-21, no. 4, pp. 466–470, 1972.

Holder D., *Electrical Impedance Tomography: methods, history, and applications*, ser. Medical physics and biomedical engineering. Institute of Physics Pub., 2005, ISBN: 9780750309523.

Holt D. R., "Periodic electromagnetic fields for finding the propagation constant of coaxial air lines with surface roughness and ohmic wall loss," *Metrologia*, vol. 46, pp. 167–177, 2009.

Homan D. N., "Application of coaxial chokes to ac bridge circuits," *J. Res. Nat. Bur. Stand.*, vol. 72C, no. 2, pp. 161–165, Apr.–Jun. 1968.

Horsky J. and Horska J., "Simulated inductance standard," in *Conference on Precision Electromagnetic Measurements CPEM 2002 Digest*, Ottawa, Canada, pp. 188-189, Jun. 16-21, 2002.

Ide J. P., "Traceability for radio frequency coaxial line standards," National Physical Laboratory (NPL), Teddington, UK, Electrical Science Report ES 114, 1992.

IEEE 1057: "IEEE standard for digitizing waveform recorders," *IEEE Std 1057-2007 (Revision of IEEE 1057-1994)*, pp. 1–142, Apr. 18, 2008, ISBN: 9780738153513.

IEEE 1241: "IEEE standard for terminology and test methods for analog-to-digital converters," *IEEE Std 1241-2000*, 2000, ISBN: 0738127248.

Inglis B. D., "Frequency dependence of electrode surface effects in parallel-plate capacitors," *IEEE Trans. Instr. Meas.*, vol. 24, no. 2, pp. 133–150, Jun. 1975.

Jackson J. D., *Classical Electrodynamics*, 2nd ed. USA: John Wiley & Sons, 1975, ISBN 0-471-43132-X.

Jackson J. D., "A curious and useful theorem in two-dimensional electrostatics," *Am. J. Phys.*, vol. 67, no. 2, pp. 107–115, 1999.

Janssen T. J. B. M., Tzalenchuk A., Yakimova R., Kubatkin S., Lara-Avila S., Kopylov S., and Fal'ko V. I., "Anomalously strong pinning of the filling factor $\nu = 2$ in epitaxial graphene," *Phys. Rev. B*, vol. 83, p. 233402, Jun. 2011.

Jeanneret B. and Benz S. P., "Application of the Josephson effect in electrical metrology," *Eur. Phys. J. - Special Topics*, vol. 172, pp. 181–206, 2009.

Jeckelmann B. and Jeanneret B., "The quantum Hall effect as an electrical resistance standard," *Meas. Sci. Technol.*, vol. 14, no. 8, p. 1229–1236, 2003.

Jenkins B. D., *Introduction to instrument-transformers*. London: Newnes, 1967.

Johnson H. and Small G., "Emulation of three terminal standard inductors with tee networks," in *Conference on Precision Electromagnetic Measurements CPEM 1998 Digest*, Washington, USA, pp. 44–45, 6-10 Jul. 1998.

Johnson J. B., "Bemerkung zur Bestimmung des elektrischen Elementarquantums aus dem Schroteffekt," *Ann. Phys.*, vol. 372, no. 2, pp. 154–156, 1922.

Johnson J. B., "Thermal agitation of electricity in conductors," *Phys. Rev.*, vol. 32, pp. 97–109, Jul. 1928.

Jones R. N., "A technique for extrapolating the 1 kc values of secondary capacitance standards to higher frequencies," NBS, Tech. Note 201, Nov. 1963.

Jones R. N., "Evaluation of three-terminal and four-terminal pair capacitors at high frequencies," NBS, Tech. Note 1024, Sep. 1980, 15 pp.

Jonscher A. K., "The universal dielectric response," *Nature*, vol. 267, pp. 673–679, Jun. 23, 1977.

Jonscher A. K., "The universal dielectric response and its physical significance," *IEEE Trans. Electrical Insul.*, vol. 27, no. 3, pp. 407–423, Jun. 1992.

Kaatze U. and Feldman Y., "Broadband dielectric spectrometry of liquids and biosystems," *Meas. Sci. Technol.*, vol. 17, no. 2, p. R17-R35, 2006.

Kalinowski J., "An inductance realization using two operational amplifiers," *IEEE Proc.*, vol. 56, no. 9, pp. 1636–1637, Sep. 1968.

Keller M. W., "Standards of current and capacitance based on single-electron tunneling devices," in *Proceeding of the 2000 International School of Physics "Enrico Fermi," course CXLVI: Recent Advances in Metrology and Fundamental Constants*, Quinn T. J., Leschiutta S., and Tavella P., Eds. IOS Press, 2001, pp. 291–313, ISBN: 1586031678.

Keller M. W., "Current status of the quantum metrology triangle," *Metrologia*, vol. 45, no. 1, p. 102–109, 2008.

Keller M. W., Martinis J. M., Zimmerman N. M., and Steinbach A. H., "Accuracy of electron counting using a 7-junction electron pump," *Appl. Phys. Lett.*, vol. 69, no. 12, pp. 1804–1806, 1996.

Keller M. W., Eichenberger A. L., Martinis J. M., and Zimmerman N. M., "A capacitance standard based on counting electrons," *Science*, vol. 285, no. 5434, pp. 1706–1709, Sep. 1999.

Keller M. W., Zimmerman N. M., and Eichenberger A. L., "Uncertainty budget for the NIST electron counting capacitance standard, ECCS-1," *Metrologia*, vol. 44, no. 6, p. 505–512, 2007.

Kester W., *The data conversion handbook.* Newnes, 2005, online at www.analog.com.

Kester W., "Taking the mystery out of the infamous formula, "SNR = 6.02N + 1.76dB," and why you should care," http://www.analog.com/static/imported-files/tutorials/MT-001.pdf, 2009.

Kibble B. P. and Schurr J., "A novel double-shielding technique for ac quantum Hall measurement," *Metrologia*, vol. 45, no. 5, p. L25, 2008.

Kibble B. P., "Four terminal-pair to anything else!" in *IEE Colloquium on interconnections from DC to Microwaves (Ref. No. 1999/019)*, pp. 6/1–6/6, 1999.

Kibble B. P. and Rayner G. H., *Coaxial AC Bridges.* Bristol, UK: Adam Hilger Ltd, 1984, ISBN: 0852743890.

Kim H. J. and Semenov Y., "Simultaneous comparison of impedance standards by means of commercial digital LCR-meters," in *Conference on Precision Electromagnetic Measurements CPEM 2008 Digest*, Broomfield, CO, USA, pp. 340–341, Jun. 8-13, 2008.

Kim H. J., Lee R. D., and Semenov Y. P., "Resistance standards with calculable frequency dependence for frequencies up to 1 MHz," in *Conference on Precision Electromagnetic Measurements CPEM 2006 Digest*, Torino, Italy, pp. 550–551, July 9-14, 2006.

Kitchin C. and Counts L., "RMS to DC conversion application guide," Analog Devices Inc., Tech. Rep., 1986, 2nd ed.

Klionsky M. D., Semenov Y. P., and Moodley A., "Determination of the frequency dependence of capacitance standards with ceramic dielectrics," in *Conference on Precision Electromagnetic Measurements CPEM 2002 Digest*, Ottawa, Canada, pp. 346–347, Jun. 16-21, 2002.

von Klitzing K., Dorda G., and Pepper M., "New method for high-accuracy determination of the fine-structure constant based on quantized Hall resistance," *Phys. Rev. Lett.*, vol. 45, no. 6, pp. 494–497, Aug. 1980.

von Klitzing K., "The quantized Hall effect," *Rev. Mod. Phys.*, vol. 58, no. 3, pp. 519–531, Jul. 1986.

Koffman A. D., Avramov-Zamurovic S., Waltrip B. C., and Oldham N. M., "Uncertainty analysis for four-terminal-pair capacitance and dissipation factor characterization at 1 and 10 MHz," *IEEE Trans. Instr. Meas.*, vol. 49, no. 2, pp. 346–348, Apr. 2000.

Kollar I. and Blair J. J., "Improved determination of the best fitting sine wave in adc testing," *IEEE Trans. Instr. Meas.*, vol. 54, no. 5, pp. 1978–1983, Oct. 2005.

Kossel M., Leuchtmann P., and Rufenacht J., "Traceable correction method for complex reflection coefficient using calculable air line impedance standards," *IEEE Trans. Instr. Meas.*, vol. 53, no. 2, pp. 398–405, Apr. 2004.

Kouwenhoven, W. B. and Lotz E. L., "Absolute power factor of air capacitors," *AIEE Trans.*, vol. 57, no. 12, pp. 766–773, Dec. 1938.

Kouwenhoven W. B. and Lemmon C. L., "Phase defect angle of an air capacitor," *AIEE Trans.*, vol. 49, no. 3, pp. 952–958, Jul. 1930.

Kramers H. A., "La diffusion de la lumière par les atomes," in *Atti del Congresso Internazionale dei Fisici*, Como, pp. 545–557, 1927.

Krönig R. L., "On the theory of dispersion of X-rays," *J. Opt. Soc. Am.*, vol. 12, pp. 547–557, 1926.

Kucera J., Vollmer E., Schurr J., and Bohacek J., "Calculable resistors of coaxial design," *Meas. Sci. Technol.*, vol. 20, no. 9, p. 095104, 2009.

Kugelstadt T., *Active Filter Design Techniques*, ser. Advanced Analog Products. Texas Instruments, Aug. 2002, vol. Op-amps for everyone, no. SLOD006B, ch. 16, pp. 1–63.

Kuperman A., Tapuchi S., Makarenko S., and Suissa U., "Capacitance-increase method," *IEEE Trans. Instr. Meas.*, vol. 59, no. 4, pp. 832–839, 2010.

Kurokawa K., "Power waves and the scattering matrix," *IEEE Trans. Microwave Theor. Tech.*, vol. 13, no. 2, pp. 194–202, Mar. 1965.

Kyle U. G., Bosaeus I., Lorenzo A. D. D., Deurenberg P., Elia M., Gomez J. M., Heitmanng B. L., Kent-Smith L., Melchior J.-C., Pirlich M., Scharfetterk H., Schols A. M., Pichard C., "Bioelectrical impedance analysis – part I: review of principles and methods," *Clinical Nutrition*, vol. 23, pp. 1226–1243, 2004.

Kyle U. G., Bosaeus I., Lorenzo A. D. D., Deurenberg P., Elia M., Gomez J. M., Heitmanng B. L., Kent-Smith L., Melchior J.-C., Pirlich M., Scharfetterk H., Schols A. M., Pichard C., "Bioelectrical impedance analysis – part II: review of utilization in clinical practice," *Clinical Nutrition*, vol. 23, pp. 1430–1453, 2004.

Lacquaniti V., De Leo N., Fretto M., Sosso A., Mller F., and Kohlmann J., "1 V programmable voltage standards based on SNIS Josephson junction series arrays," *Supercond. Sci. Technol.*, vol. 24, no. 4, p. 045004, 2011.

Lampard D. G., "A new theorem in electrostatics with applications to calculable standards of capacitance," *Proc. IEE C: Monographs*, vol. 104, no. 6, pp. 271–280, 1957.

Lampard D. and Cutkosky R., "Some results on the cross-capacitances per unit length of cylindrical three-terminal capacitors with thin dielectric films on their electrodes," *Proc. IEE C: Monographs*, vol. 107, no. 11, pp. 112–119, 1960.

Lamson H. W., "A new series of standard inductors," *General Radio Experim.*, vol. 27, no. 6, pp. 1–4, Nov. 1952.

Landau L. D. and Lifshitz E. M., *Electrodynamics of Continuous Media*, ser. Course of theoretical physics. UK: Pergamon Press, 1960, vol. 8, p. 256.

Lang D. V., "Deep-level transient spectroscopy: A new method to characterize traps in semiconductors," *J. Appl. Phys.*, vol. 45, no. 7, pp. 3023–3032, 1974.

Lee J., Schurr J., Nissilä J., Palafox L., and Behr R., "The Josephson two-terminal-pair impedance bridge," *Metrologia*, vol. 47, no. 4, p. 453–459, 2010.

Lee J., Schurr J., Nissilä J., Palafox L., Behr R., and Kibble B. P., "Programmable Josephson arrays for impedance measurements," *IEEE Trans. Instr. Meas.*, vol. 60, no. 7, pp. 2596–2601, Jul. 2011.

Leuchtmann P. and Rufenacht J., "On the calculation of the electrical properties of precision coaxial lines," *IEEE Trans. Instr. Meas.*, vol. 53, no. 2, pp. 392–397, Apr. 2004.

Likharev K. and Zorin A., "Theory of the Bloch-wave oscillations in small Josephson junctions," *J. Low Temp. Phys.*, vol. 59, pp. 347–382, 1985.

Likharev K., "Single-electron devices and their applications," *IEEE Proc.*, vol. 87, no. 4, pp. 606–632, 1999.

Linckh H. and Brasack F., "Eine Methode zur Bestimmung des Widerstandswertes aus der Induktivität," *Metrologia*, vol. 4, pp. 94–101, 1968.

Long Y.-T., Li C.-Z., Kraatz H.-B., and Lee J. S., "AC impedance spectroscopy of native DNA and M-DNA," *Biophys J.*, vol. 84, no. 5, pp. 3218–3225, May 2003.

LT1088: "Wideband RMS-DC converter building block," Linear Technology datasheet.

Makow D. and Campbell J. B., "Circular four electrode capacitors for capacitance standards," *Metrologia*, vol. 8, pp. 148–155, 1972.

Marzetta L. A., "An evaluation of the three-voltmeter method for AC power measurement," *IEEE Trans. Instr. Meas.*, vol. 21, pp. 353–357, 1972.

Mason W., Hewitt W., and Wick R., "Hall effect modulators and "gyrators" employing magnetic field independent orientations in germanium," *J. Appl. Phys.*, vol. 24, no. 2, p. 166, 1953.

Matthews E., "The use of scattering matrices in microwave circuits," *IRE Trans. Microwave Theory Tech.*, vol. 3, no. 3, pp. 21–26, Apr. 1955.

McGregor M. C., Hersh J. F., Cutkosky R. D., Harris F. K., and Kotter F. R., "New apparatus at the National Bureau of Standards for absolute capacitance measurement," *IRE Trans. Instrum.*, vol. I-7, no. 3, pp. 253–261, Dec. 1958.

Melcher J., "Systematic errors of long thin coaxial cables for the connection of four-terminal-pair devices," *IEEE Trans. Instr. Meas.* vol. 45, no. 1, pp, 339–341 (1996).

Menard D. and Yelon A., "Theory of longitudinal magnetoimpedance in wires," *J. Appl. Phys.*, vol. 88, no. 1, pp. 379–393, 2000.

Millea A. and Ilie P., "A class of double-balance quadrature bridges for the intercomparison of three-terminal resistance, inductance and capacitance standards," *Metrologia*, vol. 5, no. 1, pp. 14–20, 1969.

Mitsuo K., Suzuki K., and Yamazaki A., "A practical system to evaluate the nonlinearity of four-terminal-pair (4TP) LCR meter," in *NCSL Conf. Proc.*, Saint Paul, Minnesota, p. 7F, July 29–Aug. 2007.

Moon C. and Sparks C. M., "Standards for low value of direct capacitance," *J. Res. Nat. Bur. Std.*, vol. 41, pp. 497–507, Nov. 1948, research paper RP1935.

Moore W. J. M. and Basu S. K., "A direct-reading current-comparator bridge for scaling four-terminal impedances at audio frequencies," *IEEE Trans. Instr. Meas.*, vol. IM-15, no. 4, pp. 253–259, Dec. 1966.

Moore W. J. M. and Miljanic P. N., *The Current Comparator*, ser. IEE Electrical Measurement Series. London, UK: Peter Peregrinus Ltd, 1988, vol. 4, ISBN 0863411126.

Muciek A., "Digital impedance bridge based on a two-phase generator," *IEEE Trans. Instr. Meas.*, vol. 46, no. 2, pp. 467–470, Apr. 1997.

Muciek A. and Cabiati F., "Analysis of a three-voltmeter measurement method designed for low-frequency impedance comparisons," *Metrology and Meas. Systems*, vol. 13, pp. 19–33, 2006.

Nakamura Y., Fukushima A., Sakamoto Y., Endo T., and Small G., "A multifrequency quadrature bridge for realization of the capacitance standard at ETL," *IEEE Trans. Instr. Meas.*, vol. 48, no. 2, pp. 351–355, Apr. 1999.

Nakase T., "Precision 1 MHz capacitance standards using loss-free-type three-terminal capacitors," in *Conference on Precision Electromagnetic Measurements CPEM 1988 Digest*, Tsukuba, Japan, pp. 51–52, Jun. 7–10, 1988.

new SI brochure: ""On the possible Future revision of the SI", BIPM. [Online]. Available: http://www.bipm.org/en/si/new_si/

Novoselov K. S., Jiang Z., Zhang Y., Morozov S. V., Stormer H. L., Zeitler U., Maan J. C., Boebinger G. S., Kim P., and Geim A. K., "Room-temperature quantum Hall effect in graphene," *Science*, vol. 315, no. 5817, p. 1379, 2007.

Nyquist H., "Thermal agitation of electric charge in conductors," *Phys. Rev.*, vol. 32, pp. 110–113, Jul. 1928.

Oldham N., "Overview of bioelectrical impedance analyzers," *Am. J. of Clin. Nutr.*, vol. 64, pp. 405S–412S, Apr. 1996.

Oldham N. and Booker S., "Programmable impedance transfer standard to support LCR meters," in *Instrumentation and Measurement Technology Conference IMTC 1994 Proc.*, Hamamatsu, Japan, pp. 929–930, 10–12, May 1994.

Overney F. and Jeanneret B., "Realization of an inductance scale traceable to the quantum Hall effect using an automated synchronous sampling system," *Metrologia*, vol. 47, no. 6, p. 690–698, 2010.

Overney F. and Jeanneret B., "RLC bridge based on an automated synchronous sampling system," *IEEE Trans. Instr. Meas.*, vol. 60, no. 7, pp. 2393–2398, Jul. 2011.

Overney F., Jeanneret B., and Furlan M., "A tunable vacuum-gap cryogenic coaxial capacitor," *IEEE Trans. Instr. Meas.*, vol. 49, no. 6, pp. 1326–1330, Dec. 2000.

Overney F., Jeanneret B., Jeckelmann B., Wood B. M., and Schurr J., "The quantized Hall resistance: towards a primary standard of impedance," *Metrologia*, vol. 43, no. 5, p. 409–413, 2006.

Overney F., Jeanneret B., and Mortara A., "A synchronous sampling system for high precision AC measurements," in *Conference on Precision Electromagnetic Measurements CPEM 2008 Digest*, Broomfield, CO, USA, pp. 596–597, June 8–13, 2008.

Özkan T., Gülmez G., Turhan E., and Gülmez Y., "Four-terminal-pair capacitance characterization at frequencies up to 30 MHz using resonance frequencies," *Meas. Sci. Technol.*, vol. 18, no. 11, p. 3496–3500, 2007.

van der Pauw L. J., "A method of measuring specific resistivity and Hall effects of discs of arbitrary shape," *Philips Res. Repts*, vol. 13, no. 1, R334, pp. 1–9, Feb. 1958.

Pease B., "What's all this transimpedance amplifier stuff, anyhow?" *Electronic Design*, vol. 77, Jan. 8, 2001.

Percival D. and Walden A., *Spectral Analysis for Physical Applications: Multitaper and Conventional Univariate Techniques*. Cambridge University Press, Jun 1993, ISBN: 9780521435413.

Peterson C. W. and Knight B. W., "Causality calculations in the time domain: an efficient alternative to the Kramers-Krönig method," *J. Opt. Soc. Am.*, vol. 63, no. 10, pp. 1238–1242, Oct. 1973.

Pettinelli E., Cereti A., Galli A., and Bella F., "Time domain reflectrometry: calibration techniques for accurate measurement of the dielectric properties of various materials," *Rev. Sci. Instrum.*, vol. 73, no. 10, pp. 3553–3562, 2002.

Picotto G. B. and Pisani M., "A sample scanning system with nanometric accuracy for quantitative SPM measurements," *Ultramicroscopy*, vol. 86, no. 1–2, pp. 247–254, 2001.

Piquemal F., Devoille L., Feltin N., and Steck B., "Single charge transport standard and quantum-metrological triangle experiments," in *Proceeding of the 2006 International School of Physics "Enrico Fermi," course CLXVI: Metrology and Fundamental Constants*, Hänsch T. W., Leschiutta S., and Wallard A. J., Eds. IOS Press, pp. 181–210, 2007.

Pogliano U., "Analysis of the effect of irregularities on the electrodes of the calculable cross-capacitor," *Eur. Trans. Telecomm.*, vol. 1, no. 4, pp. 471–476, 1990.

Pogliano U., *Determinazione della costante h/e^2 mediante il condensatore calcolabile: esperimenti e studi preliminari.* CLUP Cittastudi, Nov. 1991, ISBN 8825100167.

Pogliano U. and Bosco G., "Tests on the IEN electrometric AC-DC transfer standard," *IEEE Trans. Instr. Meas.*, vol. 51, no. 1, pp. 78–81, Feb. 2002.

Pogliano U., Boella G., and Serazio D., "An investigation over superconductive inductive and capacitive components for high-precision measurements," *IEEE Trans. Instr. Meas.*, vol. 56, no. 4, pp. 1391–1395, Aug. 2007.

Pogliano U., Trinchera B., Bosco G., and Serazio D., "Dual transformer for power measurements in the audio-frequency band," *IEEE Trans. Instr. Meas.*, vol. 60, no. 7, pp. 2223–2228, Jul. 2011.

Pothier H., Lafarge P., Urbina C., Esteve D., and Devoret M. H., "Single-electron pump based on charging effects," *Europhys. Lett.*, vol. 17, no. 3, p. 249–254, 1992.

Pozar D. M., *Microwave Engineering*, 3rd ed. USA: J. Wiley & Sons., Inc., 2005, ISBN 0471448788.

Price J. C., "Frequency dependence of glass encapsulated electrometer resistors," *Electr. Lett.*, vol. 38, no. 9, pp. 413–415, Apr. 2002.

Primdahl F., "The fluxgate mechanism, part I: The gating curves of parallel and orthogonal fluxgates," *IEEE Trans. Magnetics*, vol. MAG-6, no. 2, pp. 376–383, Jun. 1970.

Primdahl F., "The fluxgate magnetometer," *J. Phys. E: Sci. Instrum.*, vol. 12, no. 4, p. 241–253, 1979.

Ramm G., "Impedance measuring device based on an AC potentiometer," *IEEE Trans. Instr. Meas.*, vol. IM-34, no. 2, pp. 341–344, Jun. 1985.

Ramm G. and Moser H., "From the calculable AC resistor to capacitor dissipation factor determination on the basics of time constant," *IEEE Trans. Instr. Meas.*, vol. 50, no. 2, pp. 286–289, Apr. 2001.

Ramm G. and Moser H., "New multifrequency method for the determination of the dissipation factor of capacitors and of the time constant of resistors," *IEEE Trans. Instr. Meas.*, vol. 54, no. 2, pp. 521–524, Apr. 2005.

Ramos P. M., Fonseca da Silva M., and Cruz Serra A., "Low-frequency impedance measurement using sine fitting," *Measurement*, vol. 35, pp. 89–96, 2004.

Randles J. E., "Kinetics of rapid electrode reactions," *Discussions of the Faraday Society*, vol. 1, pp. 11–19, 1947.

Rayner G. H., "The time-constant of carbon composition resistors," *British J. Appl. Phys.*, vol. 9, no. 6, p. 240–242, 1958.

Rayner G. H., Kibble B. P., and Swan M. J., "On obtaining the henry from the farad," NPL Report, Tech. Rep. DES 63, Mar. 1980, ISSN 01437305.

Ricketts B. W., "Audiofrequency losses in a superconducting inductor," *J. Phys. E: Sci. Instrum.*, vol. 9, no. 3, p. 179–181, 1976.

Ricketts B. W., "Capacitance bridge null detector with a superconducting inductor," *J. Phys. E: Sci. Instrum.*, vol. 11, no. 7, p. 635–638, 1978.

Ricketts B., Fiander J., Johnson H., and Small G., "Four-port ac quantized Hall resistance measurements," *IEEE Trans. Instr. Meas.*, vol. 52, no. 2, pp. 579–583, Apr. 2003.

Rietveld G., Koijmans C. V., Henderson L. C. A., Hall M. J., Harmon S., Warnecke P., and Schumacher B., "DC conductivity measurements in the Van der Pauw geometry," *IEEE Trans. Instr. Meas.*, vol. 52, no. 2, pp. 449–453, Apr. 2003.

Riordan R., "Simulated inductors using differential amplifiers," *Electronics Letters*, vol. 3, no. 2, pp. 50–51, Feb. 1967.

Ripka P., Ed., *Magnetic Sensors and Magnetometers*, ser. Remote Sensing Library. Artech House, 2001, ISBN: 9781580530576.

Rumiantsev A. and Ridler N., "VNA calibration," *IEEE Microwave Mag.*, vol. 9, no. 3, pp. 86–99, Jun. 2008.

Schmidt J. W. and Moldover M. R., "Dielectric permittivity of eight gases measured with cross capacitors," *Int. J. Thermophys.*, vol. 24, no. 2, pp. 375–403, 2003.

Schottky W., "Über spontane Stromschwankungen in verschiedenen Elektrizitäteleitern," *Ann. Phys.*, vol. 362, no. 23, pp. 541–567, 1918.

Schottky W., "Small-shot effect and flicker effect," *Phys. Rev.*, vol. 28, pp. 74–103, 1926.

Schurr J., Ahlers F. J., Hein G., and Pierz K., "The ac quantum Hall effect as a primary standard of impedance," *Metrologia*, vol. 44, no. 1, pp. 15–23, 2007.

Schurr J., Bürkel V., and Kibble B. P., "Realizing the farad from two ac quantum Hall resistances," *Metrologia*, vol. 46, no. 6, p. 619–628, 2009.

Semyonov Y. P., Klebanov I., Lee R. D., and Kim H. J., "Bifilar AC-DC resistor using a microwire," *IEEE Trans. Instr. Meas*, vol. 46, pp. 333–336, Apr. 1997.

Sharpe G. E. and Spain G., "On the solution of networks by means of the equicofactor matrix," *IRE Trans. Circuit Theory*, vol. 7, no. 3, pp. 230–239, 1960.

Shekel J., "Voltage reference node – its transformations in nodal analysis," *Wireless Engineer*, vol. 31, pp. 6–10, 1954.

Shields J. Q., "Phase angle characteristics of cross capacitors," *IEEE Trans. Instr. Meas.*, vol. IM-21, pp. 365–368, 1972.

Shields J. Q., "Measurement of four-pair admittances with two-pair bridges," *IEEE Trans. Instr. Meas.*, vol. IM-23, no. 4, pp. 345–352, Dec. 1974.

SI brochure: "The International System of Units (SI)," Organisation Intergouvernementale de la Convention du Mètre, 2006. [Online]. Available: http://www.bipm.org/en/si/si_brochure/general.html

Simonson P. and Rydler K.-E., "Loading errors in low voltage AC measurements," in *Conference on Precision Electromagnetic Measurements CPEM 1996 Digest*, Braunschweig, Germany, pp. 572–573, 17–20 Jun., 1996.

Small G. and Fiander J., "Fabrication and measurement of the main electrodes of the NMIA-BIPM calculable cross capacitors," *IEEE Trans. Instr. Meas.*, vol. 60, no. 7, pp. 2489–2494, Jul. 2011.

Smits F., "Measurement of sheet resistivities with the four-point probe," *Bell Syst. Tech. J*, vol. 37, no. 3, pp. 711–718, 1958.

So E. and Shields J. Q., "Losses in electrode surface films in gas dielectric capacitors," *IEEE Trans. Instr. Meas.*, vol. IM-28, pp. 279–284, Dec. 1979.

Somlo P. I. and Hunter J. D., *Microwave Impedance Measurement*, ser. IEE Electrical Measurement Series.   P. Peregrinus Ltd on behalf of the IEE, 1985, ISBN:9780863410338.

Sosso A., "Derivation of an electronic equivalent of QHE devices," *IEEE Trans. Instr. Meas.*, vol. 50, no. 2, pp. 223–226, Apr. 2001.

Stitt R. M., "AC coupling instrumentation and difference amplifiers," Burr-Brown, Application Bulletin AB-008A, 1990, available online on Texas Instruments website.

Straub A., Gebs R., Habenicht H., Trunk S., Bardos R. A., Sproul A. B., and Aberle A. G., "Impedance analysis: a powerful method for the determination of the doping concentration and built-in potential of nonideal semiconductor p-n diodes," *J. Appl. Phys.*, vol. 97, p. 083703, 2005.

Striggow K. and Dankert R., "The exact theory of inductive conductivity sensors for oceanographic application," *IEEE J. Oceanic Eng.*, vol. 10, no. 2, pp. 175–179, Apr. 1985.

Strouse G. F., "Standard platinum resistance thermometer calibrations from the Ar TP to the Ag FP," National Institute for Standards and Technology, Special Publication 250-81, Jan. 2008.

Stumper U., "Uncertainty of VNA s-parameter measurement due to nonideal TRL calibration items," *IEEE Trans. Instr. Meas.*, vol. 54, pp. 676–679, Apr. 2005.

Suzuki K., "A new short-bar method for 4TP admittance standards calibration by using a modified Z-matrix expression to improve signal-to-noise ratio (s/n) for higher impedances," *IEEE Trans. Instr. Meas.*, vol. 58, no. 4, pp. 980–984, Apr. 2009.

Suzuki K., Yamazaki A., and Yokoi K., "Non-linearity evaluation method of four-terminal-pair (4TP) LCR meter," in *NCSL Conf. Proc.*, Washington D.C., p. 6A, July 29–Aug. 2001.

Suzuki K., "A new universal calibration method for four terminal-pair admittance standards," *IEEE Trans. Instr. Meas*, vol. 40, no. 2, pp. 420–422, Apr. 1991.

Suzuki K., Akoi T., and Yokoi K., "A calibration method for four terminal-pair high-frequency resistance standards," *IEEE Trans. Instr. Meas*, vol. 42, no. 2, pp. 379–384, Apr. 1993.

Svetik Z. and Lapuh R., "Evaluation of capacitance and resistance standards at higher frequencies," in *Conference on Precision Electromagnetic Measurements CPEM 2000*, Digest, Sydney, NSW, Australia, pp. 642–643, May 14–19, 2000.

Swerlein R. D., "Precision AC voltage measurements using digital sampling techniques," *HP Journal*, vol. 40, no. 2, pp. 15–21, Apr. 1989.

Sze S. M. and Ng K. K., *Physics of Semiconductor Devices*, 3rd. Ed., ser. Wiley-Interscience publication.   John Wiley and Sons, 2007, ISBN: 0471143235.

Taylor B. N. and Witt T. J., "New international electrical reference standards based on the Josephson and quantum Hall effects," *Metrologia*, vol. 26, no. 1, p. 47–62, 1989.

Tellegen B., "The gyrator, a new electric network element," *Philips Res. Rep*, vol. 3, no. 2, pp. 81–101, 1948.

Thayer G. D., "An improved equation for the radio refractive index of air," *Radio Science*, vol. 9, no. 10, pp. 803–807, 1974.

Thomas J. L., "A new design of precision resistance standards," *J. Res. Nat. Bur. Std.*, vol. 5, pp. 295–304, 1930, RP201.

Thompson A. M., "The cylindrical cross-capacitor as a calculable standard," *IEE Proc. B: Electron. and Commun. Eng.*, vol. 106, no. 27, pp. 307–310, May 1959.

Thompson A. M. and Lampard D. G., "A new theorem in electrostatics and its application to calculable standards of capacitance," *Nature*, vol. 177, p. 888, May 1956.

Thomson W., "On the measurement of electric resistance," *Proc. Royal Society London*, vol. 11, pp. 313–328, 1860.

Torosyan A. and Willson A. N., "Exact analysis of DDS spurs and SNR due to phase truncation and arbitrary phase-to-amplitude errors," in *IEEE International Frequency Control Symposium Proc.*, Vancouver, BC, Aug. 29–31, 2005.

Trinchera B., Callegaro L., and D'Elia V., "Quadrature bridge for *R-C* comparisons based on polyphase digital synthesis," *IEEE Trans. Instr. Meas.*, vol. 58, no. 1, pp. 202–206, 2009.

Turgel R. S. and Oldham N. M., "High-precision audio-frequency phase calibration standard," *IEEE Trans. Instr. Meas.*, vol. IM-27, no. 4, pp. 460–464, Dec. 1978.

Twiss R., "Nyquist's and Thevenin's theorems generalized for nonreciprocal linear networks," *J. Appl. Phys.*, vol. 26, no. 5, pp. 599–602, 1955.

VIM: "JCGM 200:2008, International vocabulary of metrology–Basic and general concepts and associated terms," Available online at www.bipm.org., 2008.

Voss R. F. and Clarke J., "Flicker (1*f*) noise: Equilibrium temperature and resistance fluctuations," *Phys. Rev. B*, vol. 13, no. 2, pp. 556–573, Jan. 1976.

Vyroubal D., "A circuit for lead resistance compensation and complex balancing of the strain-gauge bridge," *IEEE Trans. Instr. Meas.*, vol. 42, no. 1, pp. 44–48, Feb. 1993.

Wagner K. W., "Zur Messung dielektrischer Verluste mit die Wechselstrombrücke," *Elektronische Zeitschrift*, vol. 40, pp. 1001–1002, 1911.

Wakamatsu H., "A dielectric spectrometer for liquid using the electromagnetic induction method," *HP Journal*, pp. 1–10, Apr. 1997.

Walker E. A., "The determination of magnitude and phase angle of electrical quantities," *AIEE Trans.*, vol. 60, no. 8, pp. 837–839, 1941.

Walsh R., "Inductive ratio arms in alternating current bridge circuits," *Phil. Mag. 7th Series*, vol. 10, no. 62, pp. 49–70, 1930.

Waltrip B. C. and Oldham N. M., "Digital impedance bridge," *IEEE Trans. Instr. Meas.*, vol. 44, no. 2, pp. 436–439, Apr. 1995.

Waltrip B. C. and Oldham N. M., "Wideband wattmeter based on RMS voltage measurements," *IEEE Trans. Instr. Meas.*, vol. 46, no. 4, pp. 781–783, 1997.

Wang Y., "Frequency dependence of capacitance standards," *Rev. Sci. Instrum.*, vol. 74, no. 9, pp. 4212–4215, Sep. 2003.

Wang Y. and Lee L., "A digitally programmable capacitance standard," *Rev. Sci. Instrum.*, vol. 75, no. 4, pp. 1158–1160, 2004.

Warburg E., "Über das Verhalten sogenannter unpolisirbarer Electroden gegen Wechselstrom," *Ann. Phys. und Chemie*, vol. 303, no. 3, pp. 67–493, 1899.

Warshawsky I., "Multiple-bridge circuits for measurement of small changes in resistance," *Rev. Sci. Instrum.*, vol. 26, no. 7, pp. 711–715, 1955.

Wheatstone C., "An account of several new instruments and processes for determining the constants of a voltaic circuit," *Phil. Trans. Roy. Soc.*, vol. 133, pp. 303–327, 1843.

White D. R. and Benz S. P., "Constraints on a synthetic noise source for Johnson noise thermometry," *Metrologia*, vol. 45, pp. 93–101, 2008.

Wilkins F. J. and Swan M. J., "Resistors having a calculable performance with frequency," *IEE Proc.*, vol. 116, pp. 318–324, Feb. 1969.

Wilkins F. and Swan M., "Precision a.c./d.c. resistance standards," *IEE Proc.*, vol. 117, no. 4, pp. 841–849, Apr. 1970.

Williams D. F., Wang J. C. M., and Arz U., "An optimal vector network analyzer calibration algorithm," *IEEE Trans. Microw.*, vol. 51, no. 12, pp. 2391–2401, Dec. 2003.

Williams J. M., Janssen T. J. B. M., Rietveld G., and Houtzager E., "An automated cryogenic current comparator resistance ratio bridge for routine resistance measurements," *Metrologia*, vol. 47, no. 3, p. 167–174, 2010.

Witt T., "Low-frequency spectral analysis of dc nanovoltmeters and voltage reference standards," *IEEE Trans. Instr. Meas.*, vol. 46, no. 2, pp. 318–321, 1997.

Witt T., "Electrical resistance standards and the quantum Hall effect," *Rev. Sci. Instrum.*, vol. 69, p. 2823–2843, 1998.

Witt T., Reymann D., and Avrons D., "The stability of some Zener-diode-based voltage standards," *IEEE Trans. Instr. Meas.*, vol. 44, no. 2, pp. 226–229, 1995.

Wolff H. H., "Wagner-earth and other null instrument capacity neutralizing circuits," *Rev. Sci. Instrum.*, vol. 30, no. 12, pp. 1116–1122, 1959.

Yamada T., Urano C., Nishinaka H., Murayama Y., Iwasa A., Yamamori H., Sasaki H., Shoji A., and Nakamura Y., "Single-chip 10-V programmable Josephson voltage standard system based on a refrigerator and its precision evaluation," *IEEE Trans. Appl. Supercond.*, vol. 20, no. 1, pp. 21–25, Feb. 2010.

Yang L., Li Y., Griffis C. L., and Johnson M. G., "Interdigitated microelectrode (ime) impedance sensor for the detection of viable salmonella typhimurium," *Biosensors and Bioelectronics*, vol. 19, no. 10, pp. 1139–1147, 2004.

Yonekura T., "A new calibration method for impedance meters," in *Instrumentation and Measurement Technology Conference IMTC 1994 Proc.*, Hamamatsu, Japan, pp. 1004–1007 vol. 2, May 10–12, 1994.

Yonekura T. and Jansons M., "High frequency impedance analyzer," *HP Journal*, vol. 45, pp. 67–74, Oct. 1994.

Yonekura T. and Wakasugi T., "Frequency characteristics of four terminal-pair airdielectric capacitors," in *Proc. Nat. Conf. Std. Labs (NCSL) Workshop and Symp.*, Washington DC, pp. 469–483, Aug. 19-23, 1990.

Yonenaga A. and Nakamura Y., "Inductance calibration method using a commercial LCR meter," in *Conference on Precision Electromagnetic Measurements CPEM 2004 Digest*, London, England, pp. 597–598, Jun. 27-Jul. 2, 2004.

Zadeh L. A., "Multipole analysis of active networks," *IRE Trans. Circuit Theory*, vol. 4, no. 3, pp. 97–105, 1957.

Zickner G., "Eine Messbrücke für sehr kleine Kapazitäten," *Elek. Nachr-Tech.*, vol. 7, pp. 443–448, 1930.

Zimmerman N. M., "Capacitors with very low loss: cryogenic vacuum-gap capacitors," *IEEE Trans. Instr. Meas.*, vol. 45, no. 5, pp. 841–846, Oct. 1996.

Zimmerman N. M., Simonds B. J., and Wang Y., "An upper bound to the frequency dependence of the cryogenic vacuum-gap capacitor," *Metrologia*, vol. 43, no. 5, p. 383–388, 2006.

Zorin A. B., Lotkhov S. V., Zangerle H., and Niemeyer J., "Coulomb blockade and cotunneling in single electron circuits with on-chip resistors: Towards the implementation of the R pump," *J. Appl. Phys.*, vol. 88, no. 5, pp. 2665–2670, 2000.

Index